Statistics
for Microarrays

Statistics
for Microarrays

Design, Analysis and Inference

Ernst Wit and John McClure
*Department of Statistics,
University of Glasgow, UK*

John Wiley & Sons, Ltd

Other Wiley Editorial Offices

John Wiley & Sons Inc., 111 River Street, Hoboken, NJ 07030, USA

Jossey-Bass, 989 Market Street, San Francisco, CA 94103-1741, USA

Wiley-VCH Verlag GmbH, Boschstr. 12, D-69469 Weinheim, Germany

John Wiley & Sons Australia Ltd, 33 Park Road, Milton, Queensland 4064, Australia

John Wiley & Sons (Asia) Pte Ltd, 2 Clementi Loop #02-01, Jin Xing Distripark, Singapore 129809

John Wiley & Sons Canada Ltd, 22 Worcester Road, Etobicoke, Ontario, Canada M9W 1L1

Wiley also publishes its books in a variety of electronic formats. Some content that appears
in print may not be available in electronic books.

Library of Congress Cataloging-in-Publication Data

Wit, Ernst.
    Statistics for microarrays : design, analysis, and inference / Ernst Wit and John McClure.
        p. cm.
    Includes bibliographical references and index.
    ISBN 0-470-84993-2 (hbk. : alk. paper)
    1. DNA microarrays–Statistical methods. I. McClure, John D. II. Title.
    QP624.5.D726W54 2004
    629.04–dc22

                                                                                2004045909

British Library Cataloguing in Publication Data

A catalogue record for this book is available from the British Library

ISBN 0-470-84993-2

Produced from LaTeX files supplied by the author and processed by Laserwords Private Limited,
Chennai, India
Printed and bound in Great Britain by TJ International, Padstow, Cornwall
This book is printed on acid-free paper responsibly manufactured from sustainable forestry
in which at least two trees are planted for each one used for paper production.

To Cristina and Carolina, my family and friends
To Helen, my family and friends

Contents

Preface

This is an unusual statistics book in several respects. It attempts to explain the statistical methodology needed to answer several biological questions. This is in contrast to books that try to teach statistical techniques and hope that the reader can find the appropriate contexts in which to use them. Many an applied scientist has seen this hope turn into despair, not really due to a fault of his own. Statistics without context can be counter-intuitive and can mislead the reader with terms such as 'significant', 'power', 'efficient', 'consistent', 'maximum likelihood', 'non-informative prior', and so on.

We focus on a very specific scientific technique—microarrays—and a set of scientific questions related to this technique. We try to show how sound statistical methods can be used to answer these questions.

As a consequence of this approach, another unusual feature of this book crops up: several statistical techniques have been explained more than once. Rather than considering this inefficient, we believe that it strengthens confidence in the technique for those who work their way through the book from A to Z; it also makes life easier for people who like to pick and choose their way through the book.

The book is divided into two parts. Part I tries to answer a question that many had hoped could have been avoided: How do I get good data? Data, all too often, arrive on the statistician's desk after the experiment has been conducted. Although this is not necessarily wrong, it sometimes leads to the very unfortunate cases in which the statistician has to tell the scientist that the data are not useful for answering the questions of interest. Help on the issue of statistical design can make the difference between not being able to begin to answer the question of interest and at least having a shot at it. Besides ideas on creating a good design, a host of other issues, ranging from the kind of data to be extracted from the slide to the quality of the data, will be addressed in this part of the book.

Part II focuses on the relationship between the data and the biological questions of interest: what do the data actually tell me? In other words, which genes are important in what situations? For example, can gene expression profiles discriminate between benign and malignant tumours? Also, research biologists are frequently interested in exploratory questions such as which genes behave in 'similar' ways. This might give some clue to their function.

In this book, we focus on sound statistical principles and logical thinking. Without this slightly skeptical stance, there may be a danger of over-interpreting random patterns that inevitably occur in microarray data. However, there *is* a wealth of information in microarray experiments, and with modern statistical techniques on fast computers, we can begin to tap into it. The overriding goal of this book, therefore, is to make this information digestible to the working scientist.

Acknowledgements:

A book like this one does not quite spring from an authentic, muse-inspired *tabula rasa*. A lot of it is hard work and not necessarily all of it has been by us. Therefore, we would like to thank several people and organizations who have been instrumental in the coming together of this book. We would particularly like to thank Keith Vass from the Beatson Institute (University of Glasgow), who, from the very beginning, has always patiently explained the biological background to this new technology. Anna Amtmann, Nighean Barr, John Bartlett & Caroline Witton, Philip Butcher & Jason Hinds, Barry Gusterson & Torsten Stein have all been extremely kind in letting us use their microarray data and were working with us on the analysis.

We would like to thank our colleagues in the Department of Statistics at the University of Glasgow for their moral support and many a cup of coffee when we needed it most. Our work, not the coffee, has been supported financially by a Faststream grant from the EPSRC (No. GRIR67125101) and an Exploiting Genomics grant from the BBSRC (No. 17/EGM17736). This gave us the chance to visit, invite and talk to colleagues, such as Mark van der Laan (Berkeley) and Federico Stefanini (Florence), Luigi Palla (Bologna), Agostino Nobile and David Bakewell (Glasgow), who have been instrumental in helping us think about several aspects of microarray analysis. Furthermore, Raya Khanin (Glasgow) has been extremely helpful in checking the book for errors, of which there were plenty. All the remaining ones are completely our responsibility.

On a personal note, we would like to thank our parents in Schagen (The Netherlands) and Reading (UK) for their support throughout. Many thanks also to Prof. J. J. Kockelmans for his continuing interest. Finally, we feel we should acknowledge Cristina and Helen, with whose help this book would almost not have been written. Nevertheless, they have been wonderful by putting up with us. From now on, we will be home at 5 p.m. Promise.

Ernst Wit and John McClure
Glasgow
December 2003

1

Preliminaries

1.1   Using the R Computing Environment

The R package is a mathematical computer language and a multifaceted computing
environment for statistics and graphics. Although there are several such packages
available, R is one of the best and, unlike almost all others, it is *free*. It has a
GNU General Public License. All you need to do to use it is to visit the R website
(http://www.r-project.org/) and download the package from there.

Because of its mathematical diversity and its free availability, R has become a
very popular tool for analysing microarrays. It is based on the S language developed
at Bell Laboratories and therefore shares many similarities with the commercial
package S-PLUS.

Although the methods we discuss in this book can be implemented in other
packages, all the plots and analyses in the book were obtained using R. In the book,
we give information on how to implement many of the procedures. The website

> http://www.stats.gla.ac.uk/~microarray/book/

that accompanies the book has more detailed information on how to use each
procedure in R, though it assumes a working knowledge of the package. Many of
the new procedures programmed by the authors have been bundled together in the
smida library. This can be downloaded from the book's website or from the R
website. Follow the instructions on how to download the latest version of R for your
platform from one of the CRAN mirrors (see http://www.r-project.org/).

If you are not familiar with R, *Introductory Statistics with R* (Dalgaard 2002) is
a useful guide. Another excellent book is *Modern Applied Statistics with S-PLUS*
(Venables and Ripley 1999), which gives an overview of many implementations of
cutting-edge statistical techniques. Although this is written primarily for S-PLUS,
the similarities between the two packages mean that it is equally relevant for R

Statistics for Microarrays: Design, Analysis and Inference   E. Wit and J. McClure
© 2004 John Wiley & Sons, Ltd   ISBN: 0-470-84993-2

users. The book's associated MASS library is available in R. Other R implementations of microarray-related algorithms are available from the Bioconductor website (http://www.bioconductor.org/).

Note that throughout this book, whenever we refer to R commands or libraries, we use typewriter font, for example, smida. All the R functions mentioned in this book are included in the smida package, unless explicitly stated.

1.1.1 Installing smida

This book not only covers many traditional statistical techniques but also proposes new methodology and algorithms for the analysis of microarray data. Many of these algorithms have been implemented by the authors of this book in R. They have been bundled into an add-on library called smida. This library is available from the book's website, and from any of the CRAN mirrors at the R website. The library has to be installed only once on your computer.

For Windows users: the easiest way to install the library in R is as follows:

1. download smida as a zip file (ends .zip), either from the book website or from one of the CRAN mirrors, and store it in a convenient folder;

2. open R on your computer;

3. click on **Packages** drop down menu and select the **Install package(s) from local zip files...** option;

4. search for the folder you chose in the window that appears, double click on smida.zip and then click on the **Open** button.

For Unix and Linux users: An easy way to install the library in R is as follows:

1. download smida as a file (ends .tgz) file, either from the book website or from one of the CRAN mirrors, and store it in a convenient folder;

2. type

        R INSTALL smida.tgz

   in a local client window in the same directory where the smida file has been saved.

For more information on how to install packages, see the section on *R Add-On Packages* in the *Frequently Asked Questions* page of the R online help. You might need system administrator privileges to install an R library. If your computer is part of a network, you may need to ask your system administrator for help in doing this.

Other platforms users: follow the instructions in your version of R.

Installing other libraries

Some of the algorithms in smida require sub-algorithms from other libraries, such as MASS or lars. Some are pre-installed in R, such as MASS, but others are not, such as lars. A list of required libraries can be found on the book website. When you load smida (see below) it will tell you which libraries need to be installed. These libraries are available from the book's website as well as from the CRAN mirrors and can be installed in the same way as the smida library.

1.1.2   Loading smida

Once you have installed the library in R, you can then load smida by simply typing

```
library(smida)
```

inside R. You need to load a library once per session. If you need to load another library, you simply replace smida with the name of that library. Note that smida is written using version 1.8 and therefore might not work properly with older versions of R.

R is a command line–driven programme. This makes investigating data sets highly interactive and flexible. On the other hand, it does require some familiarity with the S language used by R. Even for experienced users, it is extremely helpful to begin each R session by typing

```
help.start()
```

in the R command line. It launches a HTML page in your browser that automatically lists the help files for all the libraries that you have installed in R.

1.2   Data Sets from Biological Experiments

Throughout this book, we use five data sets from original microarray experiments. They are all of a very different nature and are therefore an excellent testing ground for many statistical concepts. In the Arabidopsis experiment, the relationship between potassium starvation in the plants and gene expression of transporter and other genes is of interest. The skin cancer experiment is a comparative microarray study of gene expression in a cancerous cell line and a normal cell line. The breast cancer experiment is a clinical study of DNA amplification and deletion patterns, using microarray technology. Its aim is to study the relationship between DNA amplification patterns (rather than gene expression) and the severity of the breast cancer, as measured by several clinical indicators on the patients. The mammary gland experiment studied the relationship between gene expression in the mammary gland of mice and the developmental stage of the mice. The tuberculosis experiment is also a time-course experiment, measuring the effect of starvation stress on the gene expression of the tuberculosis bacterium.

1.2.1 Arabidopsis experiment: Anna Amtmann

Dr Amtmann is a plant scientist at the University of Glasgow. She is interested in how plants regulate nutrient uptake. Plants cannot move and therefore must be able to adjust their growth and development to an ever-changing environment. Inorganic ions in the soil (e.g. potassium K^+ and calcium Ca^{2+}) are essential components of a plant's nutrition. However, both ion supply and demand fluctuate considerably depending on water status of the soil, light, time of the day, season, developmental stage and tissue. Plants overcome this fluctuation by temporarily storing the nutrients in leaves when they are in large supply and transporting them back to essential cells when they are in short supply.

In order to study the nature, regulation and physiological integration of plant ion transport, a variety of techniques can be used. Electrophysiology measures minute electrical currents passing through individual membrane proteins. Such live recordings have led to the discovery of ion channels responsible for the uptake of essential nutrients as well as toxic ions into plants. Structure–function analysis involves cloning and heterologous expression of genes to elucidate important protein regions for particular functional activities. Finally, microarrays can reveal which genes are involved in the regulation of ion transport.

By varying the physical growth environment for the *Arabidopsis thaliana* plant, Dr Amtmann triggers nutrition transport in the plant. By comparing the expression of a control plant with a plant in which nutrition reallocation takes place, it is possible to pick up differences in the gene expressions. However, the question 'which are the genes that are responsible for the regulation of ion transport?' is in fact a bit too general. It is difficult to compare the situation of nutrition transport with no nutrition transport per se. The situation of nutrition transport can be the result of two very different situations, and different genes may be responsible for each. Either the plant has been starved and is moving nutrition from storage sites into growing tissues or there is an excess of nutrients that requires the plant to store them in, for example, the leaves. The question of interest for the microarray experiment therefore is 'which are the genes that show significant changes when the plant is put under different kinds of stress to provoke nutrient transport within the plant?' The types of stress considered are varying lengths of potassium starvation and resupply.

Dr Amtmann is part of a larger group of scientists interested in a number of different nutrients, and some of their findings are published in Maathius et al. (2003).

The main aim of the experiment was to find which transporter genes are involved when Arabidopsis is starved of potassium. A number of different genes working together are needed to move potassium around the plant, and the eventual aim is to discover the regulatory network that controls potassium homeostasis under varying conditions of external supply.

Collaborators of Dr Amtmann have found that calcium and sodium transporters were strongly affected by nutrient starvation (Maathius et al. 2003). Analysis of this particular experiment found that few known potassium transporter genes responded to the potassium stress. The results could indicate that the majority of potassium transporters in Arabidopsis are regulated post-transcriptionally rather than at the transcript level.

Also, the plants used for the experiment were mature and thus were not growing quickly. Mature plants already have stores of potassium, and therefore starvation might not have induced a lot of potassium transport. Further investigations of potassium starvation will concentrate on young Arabidopsis seedlings, where changes in potassium are likely to have greater effects.

Physical aspects of the experiment

The Arabidopsis plants used in the experiment were grown hydroponically under the same laboratory conditions, initially with no mineral starvation. Once fully grown, some plants were then given a mineral starvation treatment; others were used as controls.

For the extraction of RNA, the whole roots were used. The roots are the first point of contact with external nutrient solution and so are the first part of the plant affected by a change in the surrounding conditions. The root samples are then immediately shock-frozen in liquid nitrogen. The RNA was then extracted from the roots using Qiagen RNeasy products. Sometimes spiking controls were added. A protocol from MWG (Ebersberg, Germany) was used for hybridization, using direct labelling with Cy dyes during reverse transcription and a formamide-based hybridization buffer. ScanArray Lite was used to scan the arrays, and the image analysis was carried out using QuantArray.

The arrays used were custom-made two-channel cDNA chips (MWG) containing 50-mer probes representing 1,153 genes and 57 control sequences. Each of these 1,250 oligonucleotide probes are replicated twice on the arrays. For more information, see Maathius et al. (2003).

Organizational aspects of the experiment

Eleven two-channel arrays were used, each with one channel assigned to a treatment sample and the other to a control. In seven of the arrays, the treatments involved starving the plants for different lengths of time, ranging from 5 h to 4 d. The aim of this was to give insight into which genes were responsible, in the root, for re-allocating nutrients from storage areas (such as leaves) to essential cells. The treatment samples used for the other four arrays had potassium reintroduced for 5 or 24 h after a period of starvation of 24 or 96 h. This was done to look at the recovery process of the plant after K^+ starvation. Table 1.1 details the 11 arrays'

Table 1.1 Details of the treatment samples applied to each array used in the Arabidopsis experiment, together with the dye assignment for treatment and control.

Array	Treatment type		Repeat	Cy3 channel sample	Cy5 channel sample
	K^+ starvation (Hours)	K^+ re-addition (Hours)			
1	5	None	1	Control	Treatment
2	10	None	1	Control	Treatment
3	10	None	2	Treatment	Control
4	24	None	1	Control	Treatment
5	24	None	2	Treatment	Control
6	96	None	1	Control	Treatment
7	96	None	2	Treatment	Control
8	24	5	1	Control	Treatment
9	96	5	1	Control	Treatment
10	96	5	2	Treatment	Control
11	96	24	2	Treatment	Control

treatments and the channels to which these treatments were assigned. The control sample, also known as the *reference* sample, is Arabidopsis RNA from a plant that was not starved and was harvested at the same time as the treatment plants.

Note that repeats were not taken at the same time of day and so cannot be considered exact replicates.

1.2.2 Skin cancer experiment: Nighean Barr

Dr Nighean Barr is a researcher at the Cancer Research UK Beatson Laboratories in Glasgow. One experiment that she has carried out investigated differences between gene expressions in cancerous and normal fibroblast cells. These cells are the main constituent of connective tissue within the body and make fibres and the extracellular matrix. In the skin, these cells are susceptible to UV radiation from sunlight, and through this they can become cancerous. By finding which genes are differentially expressed in the cancerous versus normal cells, one can focus research into treatments for cancer. The data have not yet been published.

Physical aspects of the experiment

The fibroblast tissue used was created *in vitro* from two cell lines—one cancerous and one normal. From each of these two cell lines, four separate technical replicates were created. Pairs of cancerous and normal replicates were then hybridized to two-channel cDNA arrays. Each array contained 4,608 genes replicated twice.

Organizational aspects of the experiment

As the RNA comes from two cell lines, the conclusions of this experiment can only be generalized to these specific cell lines. They do not necessarily have something to say about the expression in the general population of unaffected and cancerous fibroblast cells. In this sense, these replicates are sometimes referred to as technical replicates.

The experiment is a direct comparison of cancerous fibroblast cells with normal cells using four technical replicates of each of the cell lines. Four arrays were used in the experiment. The solution hybridized to each array contains one replicate of the cancerous tissue and one replicate of the normal tissue. For two of the arrays, the normal tissue was stained with Cy3 dye and Cy5 was used for the cancerous tissue; on the other two arrays the dye assignments were swapped. The design details are given in the Table 1.2.

1.2.3 Breast cancer experiment: John Bartlett

Dr John Bartlett is a senior lecturer in the Division of Cancer Sciences and Molecular Pathology of the University of Glasgow at the city's Royal Infirmary. He is interested in finding out the key genetic determinants associated with aggressive and non-aggressive breast cancer.

Rather than looking at gene expression of RNA in tissue, he investigates differences in the genomic DNA of breast cancer patients compared to that in controls. In cancer cells, the genome can undergo changes, such as obtaining additional copies of certain genes and losing genetic material from other genes. An increase in the *gene number* is known as *gene amplification*. When fewer copies are present compared to the genome of normal cells, this is known as *gene deletion*. An established method, comparative genomic hybridization (CGH) (Kallioniemi et al. 1992), allows regions of the genome where changes occur to be identified. Until recently, only fairly large regions of the genome could be considered—around 5 to 10 million base pairs. Using microarrays for conducting CGH, Dr Bartlett can consider specific genes rather than the much larger genomic regions previously looked at.

Table 1.2 Details of the dye assignments of samples applied to each array used in the skin cancer experiment.

Array	Cy3 sample	Cy5 sample
1	Cancer	Normal
2	Normal	Cancer
3	Cancer	Normal
4	Normal	Cancer

In order to link gene amplification/deletion information to the aggressiveness of the tumours in this experiment, clinical information is available about each of the patients, such as their Nottingham prognostic index (NPI). This is used to help classify the tumours into different severity groups while controlling for non-genomic influences.

Dr Bartlett is working with Dr Caroline Witton of the University of Glasgow at the Royal Infirmary and with Steven Seelig, Walter King and Teresa Ruffalo of Vysis Inc., Downers Grove, Chicago, who produced the arrays used in the experiment.

The work has resulted in a sub-classification of breast cancers and it has suggested genes that have an effect on the aggressiveness of the cancer. The results of the experiment are detailed in Witton et al. (2002). The chips used in this experiment have since been superseded by larger chips from Vysis Inc. John Bartlett and Caroline Witton continue to work on breast cancer classification with these new arrays.

Physical aspects of the experiment

Genomic DNA from cancer patients is extracted from stored frozen tumour tissue and from female reference DNA. The protocol used for the extraction, PCR amplification, labelling and hybridization are given in Pestova et al. (2004).

The arrays contain 59 clones, each spotted three times. 57 genes are represented by these 59 clones, since two genes have both 5′ and 3′ versions included. In each of the two-channel arrays, reference female DNA is used as a control in one channel. Rather than Cy dyes, the dyes used are Alexa 488, a green dye, and Alexa 594, a red one. Further details about the image analysis and arrays used are given in Pestova et al. (2004).

Organizational aspects of the experiment

The experiment involves the genetic material from 62 breast cancer patients. To measure the gene profiles in all tumours, 62 arrays were used in the experiment. In all the cases, the control samples were coloured with Alexa 594 and the patients' samples with Alexa 488. For each patient, there is also a variety of clinical information available, including the following:

- their survival time (in years) after the tissue was removed;

- their age at diagnosis (in years);

- the size of their tumour (in mm);

- whether they died from breast cancer;

- whether they are still alive;

- the severity grade of their breast cancer: 1 (low) to 3 (high);

- their NPI score.

1.2.4 Mammary gland experiment: Gusterson group

Professor Barry Gusterson and Dr Torsten Stein of the University of Glasgow's Division of Cancer Science and Molecular Pathology are interested in the molecular mechanisms of breast cancer development. They have conducted a large microarray time-series experiment looking at changes in gene expression in healthy mammary tissue during development and pregnancy.

The interest in healthy breast morphogenesis for cancer research is partly due to a desire to understand what changes in gene expression occur in non-cancerous tissue before looking at pathological tissue. Even more importantly, the morphological changes that happen during normal development show many characteristics that are also found during breast cancer development, including invasive growth, high proliferation and tissue remodelling. However, genes that regulate these changes in healthy mammary tissue are often deregulated in cancerous tissue. The identification of these genes could therefore enable the development of treatments to target such genes in cancerous patients. The time frame considered was from puberty through adult virgin, early, mid- and late pregnancy, lactation and involution, which is the period after lactation stops and the milk secreting cells die.

Given the ethical infeasibility of using human tissue, mice were used as a model organism for all experiments. In order to find the regulatory genes for the different morphological changes, it is important to identify those genes that are only expressed at certain time points of development. However, it is known that similar processes can also occur at multiple stages, and therefore similar sets of genes can be expressed at different times. Therefore, it can be important to find the genes that are expressed in certain combinations of stages.

For example, after birth, mouse pups do not suckle immediately. During this time, the mammary gland starts to go through a process that resembles early involution. Thus, it is of interest to see what genes are up- or down-regulated during both the beginning of lactation and of involution.

As a result of this project, a number of putative regulatory genes have been identified at various developmental stages. Also, around 100 genes have been found that are specifically up-regulated during involution, many of which have been independently verified. This work has been published (Stein et al. 2004); however, further analysis will be done on this large experiment.

Physical aspects of the experiment

The whole mammary gland of a mouse in the experiment was used for each array, after removing the major lymph node from the gland. The RNA from each gland was extracted after preparation and liquid nitrogen freezing using Trizol reagents and Qiagen RNeasy-columns. Following the Affymetrix guidelines, the labelled RNA was hybridized to the chip for 16 hours at 42 °C. Afterwards, the chip was washed using an Affymetrix Fluidics station. An Agilent scanner, together with Affymetrix's MAS 5.0 software, was used for the image analysis.

The chips were Affymetrix mouse arrays containing 12,488 probe sets, representing around 8,600 genes. The number of probes in each set varied between 12 and 20. Housekeeping genes were used as controls along with 4 bacterial gene spikes. Each probe is 25 nucleotides in length. Affymetrix arrays are single-channel chips and so only one condition or tissue can be probed per array.

Organizational aspects of the experiment

For each mammary gland, tissue from a different mouse is used. In the experiment, 18 time points during the development of a female mouse were considered. Each time point was replicated three times. Altogether, 54 arrays were produced. Table 1.3 details the design of the experiment.

1.2.5   Tuberculosis experiment: BµG@S group

The bacterial microarray group at St. George's hospital in London (BµG@S), led by Professor Philip Butcher, has created a number of different types of two-channel microarrays. Each array type is suitable for a different type of bacteria. The arrays have been produced not only for their own group but also for the wider scientific community. One array type is for *mycobacterium tuberculosis* (*M. tuberculosis*), the bacterium that is the cause of tuberculosis. We consider a time-course experiment conducted on *M. tuberculosis* arrays by the senior scientist at BµG@S, Jason Hinds.

The aim of the experiment was to understand the developmental changes in gene expression of *M. tuberculosis* under stressed growth conditions, where resources

Table 1.3   Table showing the stage of growth and time point within this stage of each array in the mouse mammary gland experiment.

Arrays	State	Time point	Arrays	State	Time point
1–3	Virgin	Week 6	28–30	Pregnancy	Day 17.5
4–6	Virgin	Week 10	31–33	Lactation	Day 1
7–9	Virgin	Week 12	34–36	Lactation	Day 3
10–12	Pregnancy	Day 1	37–39	Lactation	Day 7
13–15	Pregnancy	Day 2	40–42	Involution	Day 1
16–18	Pregnancy	Day 3	43–45	Involution	Day 2
19–21	Pregnancy	Day 8.5	46–48	Involution	Day 3
22–24	Pregnancy	Day 12.5	49–51	Involution	Day 4
25–27	Pregnancy	Day 14.5	52–54	Involution	Day 20

are limited. A related experiment done by the BμG@S group has been done on the *M. tuberculosis* trcS mutant and has found a number of genes differentially expressed in this mutant compared to a reference strain (Wernisch et al. 2003).

Physical aspects of the experiment

Samples of *M. tuberculosis* strains were grown in flasks with a fixed amount of growth medium. Initially, the *M. tuberculosis* grows abundantly; when resources become scarce the bacterium shuts down many of its functions. This is thought to occur at some point before day 30, the end of the experiment.

The arrays used were two-channel chips specifically designed and made by the BμG@S group in collaboration with the National Institute for Medical Research, with funding from the Medical Research Council. The array used the H37Rv strain annotation described in Cole et al. (1998). Using PCR amplification, 3,924 target genes were spotted on the arrays. These were placed on the array in 4 × 4 arrangement of 16 sub-grids.

mRNA was harvested from samples at four different time periods, thought to represent different stages of development for the bacterium in the given medium. This signal sample was converted to cDNA labelled with either the Cy3 or Cy5 dye, mixed with a reference sample and hybridized to the array. On each of the arrays, genomic *M. tuberculosis* DNA was used as the reference and was labelled with the other dye. Commonly used reference samples are typically isolated from a single representative RNA source or pooled mixtures of RNA derived from several sources. Genomic DNA offers an alternative reference nucleic acid with a number of potential advantages, including stability, reproducibility and a potentially uniform representation of all genes, as each unique gene should have equal representation in the genome (Kim et al. 2002).

Table 1.4 Table showing the time point of each array in the tuberculosis experiment together with dye assignments for the signal (mRNA) and reference (gDNA) channels.

Time Point	Replicates	Cy3 sample	Cy5 sample
Day 6	1–3	mRNA	gDNA
Day 6	4	gDNA	mRNA
Day 14	1–3	mRNA	gDNA
Day 14	4	gDNA	mRNA
Day 20	1,2,4	mRNA	gDNA
Day 20	3	gDNA	mRNA
Day 30	1–3	mRNA	gDNA
Day 30	4	gDNA	mRNA

Organizational aspects of the experiment

The four time points investigated were 6, 14, 20 and 30 days after the bacterium's introduction to the growth medium. Four different replicate samples were used at each of the time points, with one sample being used per array. Consequently, 16 arrays were used in total.

Table 1.4 gives the details of the dye assignments used for each array. Note that, although dye assignments for the signal (mRNA) and reference (gDNA) mixtures were swapped, this was not done evenly. Three of the four replicates at each time points used Cy3 for the signal and only the remaining replicate used Cy5 for the signal.

Part I

Getting Good Data

2

Set-up of a Microarray Experiment

Designing an experiment can refer both to the physical preparations as well as to the statistical advice on the number and manner of replication of the experiment. This chapter will focus on the physical design, or *set-up*, of a microarray experiment. The experimental set-up is the proper planning of sequential and sometimes parallel activities for the smooth running of the scientific discovery process. A description of the experimental set-up is very helpful, particularly for non-biologists, for aiding the understanding of the statistical issues involved. Experimental details can make the difference between a good and a useless experiment. A good and comprehensive overview of technical and biological aspects of microarray experiments can be found in Schena (2003).

2.1 Nucleic Acids: DNA and RNA

DNA is described as a double helix. It looks like a twisted long ladder. The sides of the 'ladder' are formed by a backbone of sugar and phosphate molecules, and the 'crosspieces' consist of two nucleotide bases joined weakly in the middle by hydrogen bonds. On either side of the 'rungs' lie complementary bases. Every Adenine base (A) is flanked by a Thymine (T) base, whereas every Guanine base (G) has a Cytosine partner (C) on the other side. Therefore, the strands of the helix are each other's complement. It is this basic chemical fact of complementarity that lies at the basis of each microarray.

Microarrays have many *single* strands of a gene sequence segment attached to their surface, known as *probes*. This attachment is sometimes achieved by physically spotting them on the array, as in spotted cDNA microarrays, and some-

Statistics for Microarrays: Design, Analysis and Inference   E. Wit and J. McClure
© 2004 John Wiley & Sons, Ltd   ISBN: 0-470-84993-2

times by immobilizing them to the quartz wafer surface via hydroxylation, as in Affymetrix arrays. In the future, undoubtedly, other media will become available. The single strands are, so to speak, waiting for complementary strands to bond—*hybridize*—and stick to the surface of the array.

Ribonucleic Acid (RNA) delivers DNA's genetic message to the cytoplasm of a cell where proteins are made. Chemically speaking, RNA is similar to a single strand of DNA. The purpose of a microarray is to measure for each gene in the genome the amount of message that was broadcast through the RNA. Roughly speaking, colour-labelled RNA is applied to the microarray, and if the RNA finds its complementary sibling on the array, then it naturally binds and sticks to the array. By measuring the amount of colour emitted by the array, one can get a sense of how much RNA was produced for each gene.

The process of producing colour-labelled RNA from a biological sample, such as a cell line or a tissue of interest, is a complicated process but is worth exploring in some detail. The steps involved in the process are all potential sources of variation that could possibly obscure the quantity of interest: the amount of RNA that is produced for each gene.

2.2   Simple cDNA Spotted Microarray Experiment

In this section, we describe a microarray experiment from beginning to end for a spotted, two-channel microarray. It involves an experiment of the Arabidopsis plant, whereby the main aim was to identify genes involved in ion transport. *Arabidopsis thaliana* is a small flowering plant that is widely used as a model organism in plant biology. It has the great advantage that it grows very quickly, its genome was completely sequenced in 2000, shortly after the fruit fly and the human, and its genetic structure has a wonderful complexity that shares many homologues with animals and humans.

The genome in the nucleus of a eukaryote, such as an Arabidopsis, contains the *instructions* for the activity of a cell. These instructions are first *transcribed* into RNA and then finally *translated* into proteins using a *four letter alphabet* consisting of nucleotides. The microarray is particularly suitable for measuring the transcription levels of different genes in different cells or conditions.

The experimental set-up of a microarray experiment is a long and intricate process that involves many steps. These steps depend on protocols that are continuously adjusted, and therefore this description runs the risk of becoming slowly outdated. However, the description of the experiment serves two purposes. First, it gives the statistical reader an idea about the biology and biochemistry involved in microarray technology. Second, we hope that the reader recognizes that the statistical design has a much simpler structure, irrespective of exactly which steps are involved.

2.2.1 Growing experimental material

The Arabidopsis experiment aims at studying the effects of different lengths of potassium starvation on the expression of the genes. In order to minimize spurious gene expression variation due to other reasons, the plants should be grown under strictly controlled circumstances. Daylight, temperature and medium are all artificially controlled inside a growth chamber under strict scrutiny of the scientists.

The study compares the expressions of all putative transporter genes 5, 10, 24 and 96 h after omission of potassium. The aim is to induce a reaction of the 'transport genes' that are responsible for moving the reserves from the storage sites to the growing parts of the plant. In a parallel experiment, Arabidopsis plants are grown for 24 and 96 h in a potassium-free environment before resupplying them with potassium for 5 h. This experiment is particularly aimed at finding those genes that are responsible for moving the nutrients back into storage during a replenishing period.

Many genes in plants vary during the light/dark cycle. In order to control these cyclic daylight effects, control plants are grown under the same light conditions and harvested at the same time as the treatment plants. In the experiment, the light is artificially controlled with lamps, using a cycle of 10 h of light and 14 h of darkness.

After the treatment, the plants are harvested and divided into 'roots' and 'shoots'. The roots are defined as all the plant material below the surface, whereas the shoots are all the material above the surface. Flowers and seeds are not yet present. In order to prevent naturally occurring enzymes from destroying the RNA, the tissue is immediately shock frozen using liquid nitrogen. It is then stored at 80 °C until the experiment continues.

In all microarray experiments, it is essential that the biological material should be selected under controlled conditions. Failure to do so might increase spurious variation as a result of some uncontrolled nuisance factor.

2.2.2 Obtaining RNA

Microarrays require single-stranded strings of nucleotides to be applied to them. One can use RNA that occurs naturally single stranded. Unfortunately, RNA is very unstable and needs to be kept at low temperatures to prevent degradation. In order to obtain the RNA from the harvested plant material, one uses a pestle and mortar to grind it down to a fine mush, while keeping it at very low temperatures using liquid nitrogen. Unfortunately, organic material is full of proteins that count as impurities and need to be filtered out. A membrane is used to 'fish' out the RNA. Some contamination by proteins is unavoidable. Spectrophotometry is used at 260 and 280 nm to detect the relative ratios of RNA and proteins respectively that are present in the sample. A ratio close to two counts as clean. Repetitive cleaning is applied until such ratios have been achieved. This is done for both the treatment and the control sample.

The remaining mixture for both the treatment sample and the control can then be concentrated using spin columns or precipitation. A total of 100 μg of RNA should be obtained for both samples. This quantity is sufficient for a single microarray. Modern techniques allow one to do the experiment even with a smaller quantity of RNA.

2.2.3   Adding spiking RNA and poly-T primer

The purpose of a microarray is to measure the number of transcribed copies of a particular gene in the treatment sample as compared to the control sample. Given that these quantities are fundamentally unknown, it is very difficult to anchor the observed quantities to something fixed. It has therefore been suggested to add RNA from certain unrelated genes in fixed amounts to the RNA mixture of both samples and measure the observed levels on the microarray. This would provide some kind of fixed point from which to observe the expression battlefield. In the Arabidopsis experiment, RNA from a set of eight human genes is added to the treatment and control sample. These genes are carefully chosen to make sure that they have as little as possible in common with the plant RNA. This process of measuring fixed quantities of a certain gene is called *spiking*.

Although some microarray experiments have been performed using RNA directly, most scientists prefer to work with the more stable cDNA molecules. cDNA is the 'inverse' copy of RNA. It is produced by a little inverse copy machine called an *enzyme*. It acts by copying a T for each A, an A for each T or U, a C for each G and a G for each C. In this way, it creates the inverse image of usual RNA. The copying process can be kick-started if some RNA has already been copied. Luckily, almost all interesting plant genes have a sequence of A's at the very end of their genetic sequence. This is called the *poly-A tail* of the gene transcript. By adding in a sequence of T's to the mixture, spontaneous bindings between the string of A's in the poly-A tail and the string of T's—the *poly-T tail*—will occur. The poly-T tail is also called a *poly-T primer*, because it primes or initiates the hybridization. In the Arabidopsis experiment, besides the spiking RNA, a poly-T primer is also added at this stage.

The treatment and the control sample now fill up barely half of their respective lab tubes. Just as in adding the spiking RNA and the poly-T primer, a clean pipette is used to add in some clean, distilled water. RNA is a single strand of nucleotides, that is, effectively a string of four letters, A, U, G and C. Naturally, these letters tend to bind to T, A, C and G, respectively, when they are present. This is the principle behind microarray technology. To avoid RNA binding to itself, it is heated up to 65 °C. In this way, any self-hybridizations of the RNA that has taken place is undone. After 5 min, the tubes are quickly cooled by putting them into ice for 2 min. In this way, any rehybridization of the RNA is prevented because the temperature is too low.

2.2.4 Preparing the enzyme environment

Enzymes are *picky* copying machines. They need just the right environment to do their job. First of all, they need a little priming 'to show them the way'. For this reason the poly-T bases were added, which immediately hybridized with the poly-A tail of the RNA. Also, the medium in which the enzymes live should preferably be a mixture of salt and magnesium.

Furthermore, enzymes cannot create copies of RNA out of thin air. They need the building blocks delivered to their front door. The building blocks of cDNA are the four nucleotide bases, A, T, G and C. It is therefore essential that the experimenter adds these bases to the enzyme environment beforehand. A quantity of 25 µl of each nucleotide is added to each test tube, together with the magnesium and salt buffer.

2.2.5 Obtaining labelled cDNA

The end product of a microarray experiment is an image with gene spots of varying intensity for each of the treatment and the control samples. A smart way of making genes 'visible' has been developed by adding a dye to the cDNA so that the amount of cDNA that sticks to the microarray slide could be measured via an optical scanner. In order to build cDNA, the enzyme needs the nucleotide building blocks, A, T, G and C. Rather than adding 25 µl of plain C nucleotides, the ingenious idea is to add 23 µl of C's that have a dye molecule attached to them, as well as 2 µl of plain C's. Each time the enzyme needs a C to 'copy' a G, it will most likely use one with a dye molecule attached. Therefore, the number of dye molecules present in the cDNA is proportional to the number of G's in the RNA, which is roughly proportional to the number of transcribed copies of the gene, as well as the length of the transcript. Two different dyes—Cy3 and Cy5—are used to distinguish treatment and control samples.

The reverse transcription can finally start when a viral enzyme—a *reverse transcriptase*—is added. The enzyme is stored at low temperatures because it degrades quickly at higher temperatures. It performs its best RNA copying activity, however, when it is put in a 42 °C environment. The control and treatment samples are brought up to this temperature immediately prior to adding the enzyme. Then the enzyme immediately starts its job: wherever it finds a poly-A tail that has hybridized with a poly-T primer, it continues from there to copy the other bases in a reversed way. This process is summarized in Figure 2.1.

If enzyme degradation goes faster than normal, it is sometimes advisable to add some additional enzyme after one hour. Otherwise, the enzyme is allowed to do its job for a total of two hours. At this time, enough cDNA is produced to be applied to the microarray.

2.2.6 Preparing cDNA mixture for hybridization

Not all the loose bases that were added into the RNA sample in Section 2.2.4 have been reverse transcribed into cDNA. These loose bases could possibly hybridize

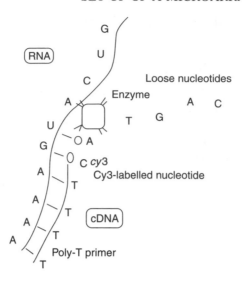

Figure 2.1   RNA preparation process for hybridization: the reverse transcriptase enzyme copies the RNA into photo-negative cDNA, which is labelled with a dye.

spuriously with the immobilized DNA on the array, and it is therefore sensible to filter them out. The mixture is passed through a membrane. The long strands of cDNA stick to the membrane, whereas the loose bases pass through. Possibly, also the RNA sticks to the membrane, but since it is not labelled, this is immaterial.

By turning the membrane the other way around, the labelled cDNA is recovered. The cDNA mixture is then dried down in a centrifuge in order to replace the liquid by a hybridization buffer. This hybridization buffer facilitates the kinetics of the actual hybridization, that is, the attachment of the cDNA produced from the sample of interest to the DNA material on the slide.

So far, all the steps have been performed for the treatment and control sample side by side. At this point, the two samples are combined in, hopefully, exactly equal quantities. The resulting mixture is then ready for hybridization to a single microarray.

The cDNA, when left at room temperature for a while may start to fold onto itself if there are complementary strands in the cDNA sequence (Figure 2.2). This would inhibit hybridization to the array, and therefore steps have to be undertaken to avoid this folding while the microarray is being prepared. By heating the cDNA mixture to 85 °C for 5 min and then shock freezing it by putting it into ice, the self-folding of the cDNA is prevented.

2.2.7   Slide hybridization

The Arabidopsis microarray is a custom array, printed for Dr Amtmann and her co-workers with approximately 1,200 sequences that represent most of the plant's

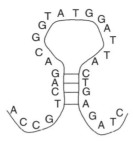

Figure 2.2   When left at room temperature, cDNA might start to fold onto itself if there are complementary strands within itself.

putative transporter genes as well as several control spots. Before applying the cDNA sample, the microarray is washed with SSC (a mixture of Sodium Chloride and Sodium Citrate) and SDS (Sodium dodecyl Sulfate). Then with a pipette the cDNA mixture containing both control and treatment samples is applied to the slide. By putting a hydrophobic coverslip on top of the mixture, the cDNA is evenly spread over the microarray and any air bubbles disappear. Finally, the slide is mounted in a hybridization chamber, fixed and put in a dark environment at 42 °C for 10 to 16 h. During this time, the actual process of interest takes place: the cDNA, which was applied to the slide, binds with complementary strands on the array. The number of matches will eventually determine the colour intensity of the scanned slide and give an indication of the amount of RNA transcript of that gene within the sample.

After the hybridization process has been completed, it is essential to remove all the labelled cDNA that did not hybridize to the slide. If sufficient care is not taken in order to remove it, the dye that is attached to that cDNA may give a spurious signal. For that purpose, the microarray is washed several times with different concentrations of SSC. Finally, the microarray is dried and is ready to be scanned.

3

Statistical Design of Microarrays

In the previous chapter, we have encountered a description of a real-life execution of a microarray experiment. The detailed experimental procedures undoubtedly show the amount of skill required to perform a microarray experiment. Moreover, the technological complexity of a microarray and the amount of preparation of the biological tissue show the progression of biology into a data-driven era.

Notwithstanding the skill of the experimenter, in such complex experiments, many sources of potential slight disturbances are possible. For instance, when performing a two-channel experiment, it is very difficult to get exactly equal amounts of both sources of cDNA. The enzyme in one tube might perform better than the enzyme in the other tube. All this has repercussions for the observed gene expressions.

Statistical design of microarray experiments aims at reducing the spurious variation by smart replication. This chapter will give several suggestions to this end. Given the high costs associated with the chips themselves, pooling true biological RNA replicates and re-scanning slides is a cost-effective way to achieve this goal. Statistical design of experiments is a form of planning of replications. Replications are repetitions of a certain experiment to decrease the uncertainty introduced in the experiment by systematic and random variations. The task of statistical design is to minimize the effect of the unwanted variations to increase the precision of the quantities of interest. Two typical constraining factors in the number of microarrays used in an experiment are the costs of the physical microarray and the amount of RNA available for performing the hybridization.

As for any kind of scientific experiment, the usual design principles also hold for microarray experiments. This means that blocking, randomization and crossing are essential to get trustworthy results. We show how these principles can be

Statistics for Microarrays: Design, Analysis and Inference   E. Wit and J. McClure
© 2004 John Wiley & Sons, Ltd   ISBN: 0-470-84993-2

implemented in several typical circumstances for both two-channel cDNA chips as well as for single-channel oligonucleotide arrays.

3.1 Sources of Variation

Any experimental scientist knows that one never gets the same results when repeating the same experiment. Whether there are several minute factors outside the control of the experimenter that cause such differences or whether there is inherent randomness in the experimental set-up, microarray experiments are subject to a lot of variation.

Some changes are systematic variations due to foreseen changes in the experiment. For instance, in a two-channel cDNA microarray experiment, the replicate might have its dyes swapped. The Cy3 and Cy5 dyes are well known to have different sensitivities (Bassett Jr et al. 1999). The first plot in Figure 3.1 shows, how in a well-replicated experiment, gene expressions with one dye (microarray 1, 2 and 3) are quite distinct from the expressions with the other dye (microarray 4). Another type of variation is introduced if the base pair length of the probes are varied across the array. Whereas Affymetrix arrays have probes with a constant length of 25 base pairs, many spotted arrays have probes with varying length. The second plot of Figure 3.1 shows clearly how there is an increasing expression pattern as a result of increasing probe lengths. These sources of variation can be accounted for.

There is also a type of unsystematic variation that behaves like chance variation, when seemingly nothing has changed in the experimental conditions. In Figure 3.1, the three points corresponding to the sample of interest measured with the same dye (6.1s, 6.2s and 6.3s) were replicates of the same condition. Chance-like variation is well known to statisticians and forms the basis of statistical theory. For example, measurement error is a common form of variability in experimental science. Statistical tools are perfectly equipped to deal with this form of variation.

Finally, bias is a type of variation that can spell disaster to any kind of experiment. Although one tries to control the experiment, keeping as many of the circumstances constant, there is always the possibility that something changed without the experimenter knowing. It could be that the room temperature during two hybridizations differed, which, unbeknownst to the experimenter, caused a subtle shift in the results. There are many potential influencing factors that are impossible to control at all. We shall show how randomization is able to guard against such systematic, but unknown variation.

We can broadly classify the variables according to distinct parts of a microarray experiment. Table 3.1 extends the table in Draghici et al. (2001) in describing the sources of variation in a typical two-channel cDNA microarray experiment. Many of these variation sources also apply to other technologies, such as Affymetrix oligo-chips.

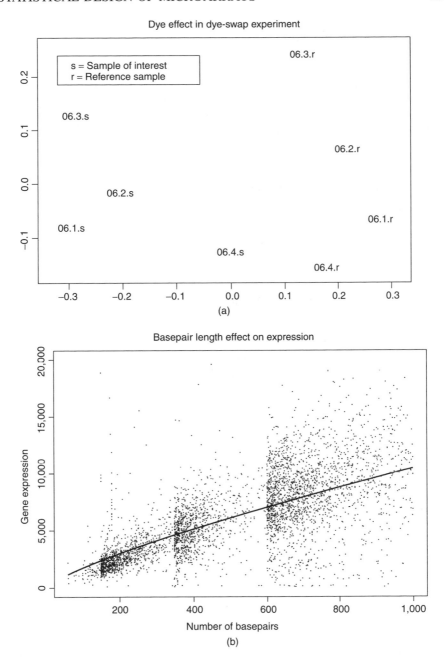

Figure 3.1 (a) A two-dimensional Sammon representation of four 4,624-dimensional two-channel microarray replications. The fourth microarray is slightly peculiar. This is the microarray in which the dyes were swapped. (b) There is a clear correlation (0.67) between the expression pattern and the base pair length of the probes.

Table 3.1   Sources of variation in a cDNA microarray experiment.

Sources of variation in the mRNA

Differences in conditions.
Differences between experimental subjects within the same covariate level.
Differences between samples from the same subject.
Variation in mRNA extraction methods from original sample.
Variations in reverse transcription.
Differences in PCR amplification.
Different labelling efficiencies.

Sources of variation in the microarray production

Print-pin anomalies.
Variations in printed probe quantities even with the same pin.
Chip batch variation (due to many sources of unknown variations).
Differences in sequence length of the immobilized DNA.
Variations in chemical probe attachment levels to the slide.

Sources of variation in the hybridization process

Different dye sensitivities.
Inequalities in the application of mRNA to the slide.
Variations in the washing efficiencies of non-hybridized mRNA off the slide.
Other differences in hybridization parameters, such as:
• temperature
• experimenter
• time of the day.

Sources of variation in the scanning

Different scanners.
Different photo-multipliers or gain.
Different spot-finding software.
Different grid alignments.

3.2 Replication

Why replicate at all? If there were no other variability in the experiment except as a result of different conditions, then there would indeed be no point in replicating an experiment. One sample from any subject in each of the conditions would be able to establish a unique expression profile for that condition. Replication comes into the picture only in order to average out or in other ways to eliminate variations that are due to sources of non-specific variation and in order to establish the significance of these results. Sometimes, although this is quite rare, biologists may be interested in the biological variation itself.

A lot of confusion exists over what should be replicated. Replication can occur on many different levels. A cursory glance at Table 3.1 shows how certain things can be kept constant while others may vary. For example, one could replicate RNA by taking a new tissue from a different subject or by taking a different PCR amplification of the RNA from the same tissue from the same subject.

In general, the correct level of replication is determined by the population about which inference is required. The only thing that is good enough to answer biological questions are the so-called biological replicates. If the final aim of an experiment is to make inference about breast tumours in general, then each sample should be a biologically independent breast tumour sample. Samples from the same breast cancer cell line are not good enough. If only such technical replicates are used, then the conclusions will necessarily be limited to the cell line from which the samples came.

3.2.1   Biological and technical replication

Replication in microarray experiments primarily involves the choice of a new biological sample to serve as a microarray target. Replicating biological samples is a labour-intensive process and often involves PCR amplification, reverse transcription and labelling. Instead, it would be much more convenient for an experimenter to prepare a whole batch of mRNA in one run, reverse transcribe it together, label it together and then apply it to different arrays. Unfortunately, this may lead to serious bias. Consider the following example.

Example 3.1 Biological replicates. Table 3.2 shows several hypothetical expressions for one gene in two cases. The first case is the situation in which the RNA

Table 3.2   Five hypothetical expressions for one particular gene with true average expression 6. The expressions from the same subject are very stable but biased, whereas the expressions from different subjects are more variable but unbiased.

Run	Same subject	Different subjects
1	5	5
2	6	8
3	5	4
4	4	6
5	5	7
mean	5	6
s.d.	0.7	1.6

comes from the same subject, whereas in the second situation, the tissue is obtained each time from a completely new subject.

Whereas expression data from the same subject is very stable, there is no way to find out if it is correct on average. Although the data from different subjects is necessarily more variable, it does, on average, converge to the right answer. By varying the subjects, the bias that each subject introduces is almost eliminated by taking the mean.

The best way to evaluate the strength of your replication scheme is by counting the *degrees of freedom*. The degrees of freedom are the number of independent samples from the populations of interest minus the number of populations you wish to compare. As a rule of thumb, the degrees of freedom should always be positive and preferably larger than at least five to be able to do statistical inference.

For example, in an experiment to compare a transgenic mouse type with a wildstrain mouse type, four mice are sampled from each strain. From each mouse, three aliquots are obtained. Finally, the 24 samples are hybridized to 12 two-channel microarrays. Despite the 12 pairs of observations, this experiment has only 6 degrees of freedom (8 mice minus 2 strains).

The actual gene expression values are the result of a particular reading of the microarray by a scanner linked to a particular image analysis programme as a result of a particular gridding procedure. Each scanner and each programme might have their peculiarities and cause a specific bias to the spot intensity values. Re-gridding and re-scanning slides by several different imaging programmes can be considered a form of technical replication, which does not increase the degrees of freedom but eliminates errors as a result of a particular scanning, gridding or programme flaw.

Often, chip designs give the experimenter the opportunity to replicate genes or ESTs on the same array. Affymetrix chips replicate each gene or ESTs between 11 and 20 times on each array. Despite these forms of replications, each chip in its entirety is exposed to one and the same labelled mRNA. Therefore, this form of replication is not biological replication and does not, therefore, increase the degrees of freedom of the experiment. Nevertheless, these technical replicates may contain useful information. Each of the technical replicates captures the variability due to measurement error and due to hybridization inequalities across the slide. By taking appropriate averages across technical replicates, bias due to mere measurement error and hybridization inequalities can be reduced.

Since eventually one is interested in finding out the average expression level of a gene under a particular condition, each new hybridization sample should preferably come from a biological replicate out of the population of subjects with the particular condition of interest. Only in this way is it safe to assume that none of the variation introduces bias. The importance of biological replication instead of technical replication has been stressed by several authors (Churchill 2002; Draghici et al. 2001; Yang and Speed 2002).

3.2.2 How many replicates?

Lee et al. (2000) derive the result that microarrays triplicates suffice for most forms of inference. This result has been often quoted (e.g. Yang and Speed 2002), if not as a gold standard then at least as a handy rule of thumb. In the original paper, Lee et al. (2000) attain this result by considering only the replicability of technical replicates on a single slide. Churchill (2002), indeed, reports that technical replicates on the same slide are highly reproducible (correlation of 0.95), but it is the reproducibility of biological replicates that is important for inference. Churchill reports that biological replicates can have a correlation as low as 0.30. Sometimes it is suggested that with three replicates it is possible to detect one anomalous observation. Besides biasing the variance estimate, this 'method' cannot serve as the basis for any formal procedure.

The two most obvious reasons why only a small number of replicates can be considered is the limited budget and perhaps the limited availability of mRNA from different biological replicates. Whereas these are valid experimental constraints, it is important to see how under these constraints it is possible to optimize the experimental design. This depends on the amount of noise present in the system and the amount of certainty needed for the conclusions.

Noise levels vary from one experimenter to the next and from one slide technology to the next. Extremely careful experimenters using a very precise platform and pooling a lot of mRNA could theoretically need only one replicate to evaluate differential expression, even though they would not have any way of confirming this result statistically. Human experimenters need replication for two purposes: (i) to eliminate some of the inevitable variability by averaging the results and (ii) to evaluate the statistical significance of their conclusions.

If microarrays were to be used for confirmatory analysis in which a high degree of certainty about the conclusions has to be achieved, then many replications would be required. Wernisch (2002) derives an approximation of the number of microarrays n needed to achieve $(1 - \alpha)$ certainty to detect a f-fold differential expression, which can be expressed as

$$n \geq \frac{\sigma^2 \log f}{z_{1-\alpha}^2},$$

where $z_{1-\alpha}$ is the $(1 - \alpha)$th quantile of the standard normal. The next section is dedicated, in part, to extending this result in order to derive a method to determine the optimal number of arrays and pooling given the variability present in the system and the level of accuracy required. Because of multiple comparisons of many genes, the statistical significance levels should be taken merely as a guide.

3.2.3   Pooling samples

Harvesting RNA can be difficult, especially when dealing with small organisms such as plants. It is often necessary to combine RNA from several subjects in the population to generate enough for application to a single microarray slide. This is one occasion in which vice happens to be a virtue. By pooling the RNA from several subjects, one effectively reduces the bias that each member of a population introduces. Especially, if one has few chips available and is truly only interested in the mean expression levels within a population and not in the variation of that expression within that population, then this approach is a relatively cheap way to maximize the accuracy of one's results. The advantages of pooling biological samples to reduce the number of microarrays are also mentioned elsewhere (Churchill 2002; Churchill and Oliver 2001). In this section, we closely follow the setting introduced by Kendziorski et al. (2003).

The literature is not uniformly in favour of pooling. It has been suggested that 'pooling samples can also have the effect of averaging out the less significant changes in expression' (Affymetrix, GeneChip Help Notes). Perhaps, in one biological sample a certain gene is up-regulated, whereas in another biological sample this is not the case. Obviously, the result of pooling is that this gene may not show any large expression change. However, this averaging effect is the whole purpose of pooling. A researcher is not interested in the particulars of the sample but in the generalizations to a larger population, such as the population of all breast cancer patients.

Although it is true that pooling prevents the estimation of the variability between individual mRNA samples, this does not preclude the construction of appropriate test statistics for determining significant differential expression. Even if the question of interest is to determine a normal *range* of values within the population of interest, then pooling can still be useful (see discussion around Equation (3.2)). We discuss a method for selecting an optimal pooling strategy, such that the variance of the test statistic of interest achieves a certain pre-specified minimum.

The idea can be summarized in a few lines. If n samples from a particular gene in one condition are expressed with a standard deviation of σ, then by mixing an independent collection of n RNA samples, the observed biological variation reduces to only $\sigma/\sqrt{n}$. For example, if each hybridization sample is made by pooling four mRNA samples of different individuals within one condition, then this reduces the biological variation by a factor of 2. It is essential that independent RNA samples be used. If the RNA sample is created by aggregating replicates that have been exposed to the same treatment simultaneously, then a smaller reduction in variance is expected. Also, it is important that pooled RNA and its constituents are applied to only one microarray. Replicate arrays should include pooled RNA from completely different RNA samples.

Optimal pooling

In this section, we shall make some practical suggestions about pooling samples if one is interested in minimizing the cost of an experiment. Put in another way, we shall try to find which replication scheme minimizes the variance given a fixed budget. We shall call this approach *optimal pooling*. The mathematical details of the approach are described in a later section.

The choice for pooling is really a decision about the trade-off between the cost of sampling RNA and the cost of a microarray. If the extraction of an RNA sample is very expensive relative to the cost of a microarray, pooling samples to reduce variation is not particularly efficient, whereas if the extraction of RNA is quite cheap, then pooling a lot of them on a single array is exactly the thing to do.

We indicate the cost of a single RNA extraction by C_s. This includes the costs of actual extraction and pre-processing. However, it does not, importantly, include the costs of the reverse transcription and labelling. These costs are included in the cost C_a of a single microarray. For dual-channel microarrays there are two possibilities. If a reference design (*cf.* reference designs in Section 3.5) is chosen, C_a is the cost of the whole array including both dyes. If both channels contain samples of interest (*cf.* loop and related designs in Section 3.5), C_a is half of this value.

The quantity σ_ϵ represents biological variation. It stands for the gene expression standard deviation for a particular gene in a particular condition if no technical and measurement error were present. The quantity σ_η, on the other hand, represents this type of technical and measurement variation.

Important result: If one wants to be certain that an f-fold difference with $1 - \alpha$ confidence will be detected, then the optimal number of arrays n_a and the optimal number of samples pooled per array n_s can be written as

$$n_a = \frac{\sigma_\eta^2 + \sigma_\epsilon \sigma_\eta \sqrt{C_s/C_a}}{\sigma_0^2} \quad \text{and} \quad n_s = \frac{\sigma_\epsilon^2 + \sigma_\epsilon \sigma_\eta \sqrt{C_a/C_s}}{\sigma_\eta^2 + \sigma_\epsilon \sigma_\eta \sqrt{C_s/C_a}}, \tag{3.1}$$

where $\sigma_0^2 = \frac{1}{2} \left(\frac{\log f}{z_{1-\alpha}} \right)^2$ and the quantity $z_{1-\alpha}$ is the $(1 - \alpha)$th quantile of the standard normal distribution.

Example 3.2 In the derivation, the values for biological variation σ_ϵ, technical variation σ_η, cost per chip C_a and cost per sample preparation C_s are assumed to be known and fixed. Whereas the first value is truly fixed, perhaps depending only on the particular gene or EST, the other three values change from lab to lab, platform to platform and certainly also over time. Continual improvement of microarrays is likely to shrink the technical variation σ_η, whereas economies of scale is likely to reduce the costs associated with the chip and with sample preparation.

Here, we deduce some reasonable current values for these constants, based on numbers from the general literature. Other estimates can be generated by simply replacing the values with one's own numbers. Kendziorski et al. (2003) suggest that $700 is a reasonable price for a microarray, including reagents, dyes and labour, whereas $50 covers for the extraction of a typical RNA sample including labour.

Churchill (2002) found that the correlation between two arrays hybridized with the same RNA is approximately 0.70, while this correlation for chips with RNA from different biological replicates is just over 0.30—approximately 0.40 in our experience. If θ is considered a random variable, namely as the expression of a randomly selected gene, then the correlations can be expressed as a combination of the quantities of interest,

$$\mathrm{Cor}(\theta + \epsilon_1 + \eta_{11}, \theta + \epsilon_1 + \eta_{12}) \approx 0.7,$$

$$\mathrm{Cor}(\theta + \epsilon_1 + \eta_{11}, \theta + \epsilon_2 + \eta_{21}) \approx 0.4,$$

where $\theta + \epsilon_i$ is the actual expression of a particular sample i and η_{ij} is the measurement error associated with the jth technical replicate of sample i.

This results in the following set of equations:

$$\frac{\sigma_a^2 + \sigma_\epsilon^2}{\sigma_a^2 + \sigma_\epsilon^2 + \sigma_\eta^2} \approx 0.7,$$

$$\frac{\sigma_a^2}{\sigma_a^2 + \sigma_\epsilon^2 + \sigma_\eta^2} \approx 0.4,$$

where σ_a is the variation on a single microarray. Although this value depends on the experimental settings and the type of chip used, we have observed array-wide standard deviations between 0.6 and 1.0 for the log expression values in several full genome, double precision (16 bit) arrays. For this example, we shall assume it is 0.8. Combining these figures with the correlations that Churchill observed and the expressions above, it is easy to derive expressions for σ_ϵ and σ_η:

$$\sigma_\epsilon = 0.6 \quad \text{and} \quad \sigma_\eta = 0.6.$$

If we would like to detect a 1.5 fold change with 90% confidence, then Equation (3.1) yields the optimal values for the number of arrays and the number of samples:

$$n_a = \frac{0.6^2 + 0.6\,0.6\sqrt{\frac{50}{700}}}{\frac{1}{2}\left(\frac{\log 1.5}{1.28}\right)^2} = 9.1 \quad \text{and} \quad n_s = \frac{0.6^2 + 0.6\,0.6\sqrt{\frac{700}{50}}}{0.6^2 + 0.6\,0.6\sqrt{\frac{50}{700}}} = 3.74.$$

This means that under these circumstances an approximately optimal pooling regime is found by applying to each microarray a pool of four biological replicates. Approximately nine microarrays are needed per condition to make sure that the required detection is obtained. Clearly, if replication improves, fewer microarrays are needed, while if the price of a microarray goes down relative to the price of a sample preparation, then more arrays will be used relative to the number of pooled samples.

Estimation of biological variation under pooling

It has been argued that 'pooling hides information about biological variance that may be informative' (http://www.med.upenn.edu/microarr/faq.htm). If a researcher happens to be interested in the *normal variation* of a particular gene under a certain condition, then this charge would be quite serious and should not be taken lightly. Although pooling clearly reduces the biological variation, it does not, however, prevent the estimation of biological variation as long as there are at least two independent pools on separate arrays.

For example, if there are n chips for a particular condition and each of the chips contains a pool of k biological replicates, then for each gene there are n observations available:

$$x_1, x_2, \ldots, x_n.$$

The variation associated with these observations is not a direct estimate of the biological variation. Instead, it is smaller due to pooling. The standard deviation of these numbers in the pooled experiment, σ_{pool}, can be written as a function of the biological standard deviation σ_{bio}:

$$\sigma_{\text{pool}} = \frac{\sigma_{\text{bio}}}{\sqrt{k}}.$$

This follows directly from standard probability theory. This can be used to estimate the biological variation from the observed variation:

$$\hat{\sigma}_{\text{bio}} = \sqrt{k}\hat{\sigma}_{\text{pool}}$$

$$= \frac{\sqrt{k}}{n-1} \sum_{i=1}^{n} (x_i - \overline{x})^2. \tag{3.2}$$

The estimate is of course only as good as the number of chips available. It does not depend on the number of biological samples in each pool.

Implementation of optimal pooling in R: op.

The function op is designed to calculate the optimal number of arrays and samples per arrays to minimize the costs while still maintaining a f-fold detection with a confidence level of 1-alpha. The user has to specify the array.cost and the sample.cost, as well as the biological variation, bio.sd, and the technical variation, measurement.sd. The defaults for these values are taken from the example in this section.

Mathematical details

While the general $1/\sqrt{n}$ rule governs the use of pooled samples, it is possible to deduce the optimal combination of pooling and replication to achieve the best possible detection of differential expression using the minimal amount of resources. Pooling samples is not without costs, and at a certain point, the reduction of biological variation as a result of pooling is outweighed by these costs. In this section, we deduce a formula for the optimal amount of pooling.

Let x_{ij} be the observed log expression of a particular gene in pool i for the jth replication of that pool to an array. We write the expression as a combination of true expression (θ), biological, between-pool variation (ϵ) and technical variation (η),

$$x_{ij} = \theta + \epsilon_i + \eta_{ij}.$$

The effect of the between-pool variation depends intrinsically on the number of samples in the pool. The more samples within a pool, the less distinguishable the pools become. Nevertheless, even if pools are mixtures of an infinite number of samples and even if two samples of the same pool are applied to two arrays, the expression of the same gene will not be the same. The reason for the difference is the result of the particular circumstances under which each microarray is hybridized. The size of this effect is independent of pool size, since it is only hybridization-specific.

The log transformation for microarray data approximately stabilizes the variances and makes gene-specific expression values approximately normal. For mathematical simplicity, we assume, therefore, that both between-pool and within-pool variation is normally distributed, according to

$$\epsilon_i \sim N\left(0, \frac{\sigma_\epsilon^2}{s_i}\right) \quad i = 1, \ldots, n_{\mathrm{a}},$$

$$\eta_{ij} \sim N(0, \sigma_\eta^2) \quad j = 1,$$

where s_i is the number of samples in pool i and n_{a} is the number of pools. Each pool is hybridized only once to a microarray. The standard deviation $\sqrt{\sigma_\epsilon^2/s_i}$ is a direct result of the $1/\sqrt{n}$ rule. The total number of samples needed for this experiment, t_{s}, can be written as

$$t_{\mathrm{s}} = \sum_{i=1}^{n_{\mathrm{a}}} s_i.$$

A natural estimate of the expression level θ is the sample mean $\bar{x}$. Other robust versions of the location parameter, such as a weighted mean or the median, are valid as well. Optimality results do not differ very much for either choice of the location parameter. Moreover, if the normality assumption is not violated and the variances

σ_ϵ^2 / s_i are all equal, then the mean is the best unbiased estimate. The variance of the mean $\bar{x}$ is given as,

$$V(\bar{x}) = V\left(\frac{1}{n_a} \sum_{i,j} [\theta + \epsilon_i + \eta_{ij}]\right)$$

$$= \frac{1}{n_a^2}\left[\sum_{i=1}^{n_a} V(\epsilon_i) + \sum_{i,j} V(\eta_{ij})\right]$$

$$= \frac{\sum_{i=1}^{n_a} 1/s_i}{n_a^2}\sigma_\epsilon^2 + \frac{1}{n_a}\sigma_\eta^2. \tag{3.3}$$

In order to minimize the variance of the estimate of the log-expression level, it is clear that increasing both the number of arrays n_a as well as increasing the number of samples in each pool s_i will reduce the overall variance. However, increasing both results in spiralling costs and some kind of balance needs to be found. We assume that there are two types of cost associated with the number of samples in each RNA pool and the number of arrays used in the experiment. Let

$$C_s = \text{cost per sample},$$

$$C_a = \text{cost per microarray}.$$

So, the optimization problem can be formulated as minimizing the costs,

$$C(s_1, \ldots, s_{n_a}) = t_s C_s + n_a C_a, \tag{3.4}$$

under the constraint of fixing the variance at a certain level σ_0^2,

$$V(\bar{x}) = \sigma_0^2. \tag{3.5}$$

From the point of view of symmetry, it is clear that the optimum is found at $n_s := s_1 = s_2 = \ldots = s_p = t_s/n_a$. This reduces the expression of the variance of the log expression in Equation (3.3) to

$$V(\bar{x}) = \frac{\sigma_\epsilon^2}{t_s} + \frac{\sigma_\eta^2}{n_a}. \tag{3.6}$$

Euler–Lagrange optimization can be used to minimize the costs under the constraint of keeping the expected variation to a preset level. We find the minimum of the objective function f,

$$f(t_s, n_a, \lambda) = t_s C_s + n_a C_a + \lambda\left(\frac{\sigma_\epsilon^2}{t_s} + \frac{\sigma_\eta^2}{n_a} - \sigma_0^2\right),$$

where λ is the Lagrange multiplier, by setting the following system of equations to zero:

$$\frac{\delta}{\delta t_s} f(t_s, n_a, \lambda) = C_s - \lambda \frac{\sigma_\epsilon^2}{t_s^2}$$

$$\frac{\delta}{\delta n_a} f(t_s, n_a, \lambda) = C_a - \lambda \frac{\sigma_\eta^2}{n_a^2}$$

$$\frac{\delta}{\delta \lambda} f(t_s, n_a, \lambda) = \frac{\sigma_\epsilon^2}{t_s} + \frac{\sigma_\eta^2}{n_a} - \sigma_0^2.$$

This system is easily solved, and it can be shown that a minimum is found at

$$t_s = \frac{\sigma_\epsilon^2 + \sigma_\epsilon \sigma_\eta \sqrt{C_a/C_s}}{\sigma_0^2}, \quad \text{and} \quad n_a = \frac{\sigma_\eta^2 + \sigma_\epsilon \sigma_\eta \sqrt{C_s/C_a}}{\sigma_0^2}. \tag{3.7}$$

From this it is clear that the solution only depends on the relative ratio of the array versus sample costs C_a/C_s and on the relative ratio of the biological and technical variability versus the prescribed total variability, σ_ϵ/σ_0 and σ_η/σ_0, respectively.

The variability constraint (3.5) is most easily interpreted and reformulated in terms of detectable fold-changes. If we assume the same error model for the log expressions y_{ij} of the same gene in another condition, then the differential expression test statistic,

$$T = \overline{x} - \overline{y} \tag{3.8}$$

has a variance of $2\sigma_0^2$. If one wishes to detect f-fold increases at a $(1 - \alpha)\%$ confidence level, then this means solving $P(T < 0) = 1 - \alpha$ where $T \sim N(\log f, 2\sigma_0^2)$, i.e.,

$$\frac{\log f}{\sqrt{2\sigma_0^2}} = z_{1-\alpha}.$$

Therefore, we can substitute the following expression for σ_0^2 in the optimum Equation (3.7):

$$\sigma_0^2 = \frac{1}{2} \left(\frac{\log f}{z_{1-\alpha}} \right)^2. \tag{3.9}$$

An argument can be made for replacing $z_{1-\alpha}$ by $z_{1-\alpha/2}$. The latter is more stringent and corresponds to a two-sided test as opposed to a one-sided test.

3.3   Design Principles

When performing a microarray experiment, one comes across many *practical* questions. In an experiment, should arrays from the same batch or from different batches be used? If one has access to several scanners, should one scan the array with one scanner or with several scanners? And, does this matter? In a cDNA experiment, should one aim for probes with the same number of base pairs? The answer to these questions are inherently statistical.

Statistical design was first introduced by Ronald A. Fisher in the 1920s for designing agricultural field trials. The relevance of these ideas when designing a scientific experiment has only increased since then. The design of agricultural field trials applies directly to microarray experiments.

In order to explain the principles of statistical design, several design concepts have to be explained. First, the terms *treatment* or *condition* stand for any attribute of primary interest in the experiment. 'Condition' is generally used to describe an immutable attribute, such as, for instance, 'wild type' or a particular cancer strain, whereas 'treatment' stands for an attribute that, in principle, can be assigned, such as, for instance, a particular sampling time point or a radiation treatment.

A statistical *unit* is an independent replicate that is subject to a condition or treatment of interest. Technically speaking, each microarray is a statistical unit. It is, however, more common (Glonek and Solomon 2004; Yang and Speed 2002) to consider each gene or EST as a statistical unit.

A *blocking factor* is a condition that is expected to have some influence on the outcome but which is not of any particular interest. It is known, for instance, that different print-pins might have different efficiencies. Other examples of blocking factors are experimenters and microarray batches.

When *crossing* factors, one assigns all possible combinations of these factors to the units. The most common cross is that of the gene factor with the dye factor, which is more commonly known as a dye-swap experiment. If factors are not properly crossed, then *confounding* can occur. Two factors are confounded when the levels of one factor are always observed with exactly the same levels of the other. For example, if the condition A is always hybridized in the Cy3 channel and condition B in the Cy5 channel, then dyes and conditions are confounded. The danger of confounding is that it makes it impossible to decide whether the observed changes are the result of one or the other factor.

3.3.1 Blocking, crossing and randomization

The principles of microarray design can be summarized by the following three general design ideas (Cobb 1998; Draghici et al. 2001):

1. **Blocking**: Subdivide the units into blocking factors, that is, groups of similar units; then assign the treatments or conditions of interest to units separately within each block.

2. **Randomization**: A chance device should be used to assign treatments or conditions to units.

3. **Crossing**: In order to compare the effects of two or more sets of conditions in the same experiment, cross them. This means that you should take the set of all possible factor combinations as your potential set of experiments.

Replication involves a careful identification of potential sources of variation and deliberately varying them while keeping only the conditions of interest fixed. If microarrays from several different batches are available, then it is important to acknowledge this explicitly by including a blocking factor. Blocking transforms unplanned, systematic bias into planned, systematic variation.

Confounding

It would be undesirable, however, to assign all the samples from one condition to one batch and all those of the other condition to another batch. In this set-up, it would be impossible to decide whether observed differences between gene expressions from the two conditions are due to these conditions or due to the different batches. This is a classical example of confounding. One can avoid confounding by making sure that the variables are appropriately crossed. For example, if two batches of microarrays are available, then half of the mRNA of each of the two conditions should be hybridized using microarrays from one batch and the other half to the arrays from the other batch.

Randomization

Even if blocking and crossing are consistently applied, the risk of bias still exists. For example, it may not seem unreasonable to hybridize within each batch the mRNA from the first condition to microarrays with an even serial number and mRNA from the second condition to arrays with an odd serial number. This seems to satisfy the requirement of properly crossing conditions within each block.

However, what if, unbeknownst to the experimenter, the chips with even serial numbers were spotted by one spotting machine and those with odd serial numbers by another. So, whereas the experimenter believes that the differential expression is the result of the two different conditions, it is in fact due to the two different spotting instruments. This may seem far-fetched, but science is riddled by such horror stories. Is there any way to protect an experiment against any such unknown systematic bias? The only way this is possible is by avoiding any kind of systematic assignment. The best non-systematic way to assign the treatments to the units in each block is by randomization. Experimenters should keep a coin in their office to assign, for example, the mRNA samples from different conditions to the different microarrays in a batch. Randomization is a key instrument in avoiding bias: one should make random rather than arbitrary decisions in the design of an experiment.

Unfortunately, not all effects can be blocked or randomized. For example, it is difficult to vary the array layout within a single experiment. Manufacturers of microarrays mass-produce chips with a fixed layout of genes or ESTs. Even the layout of custom arrays is constant for a whole batch of arrays. This could be a potential problem if certain areas of a slide are subject to systematic effects.

3.3.2 Design and normalization

In the microarray field, a whole normalization industry has appeared that traditionally used to be part of statistical design. Normalization promises to eliminate systematic artifacts from the data and to return a 'cleaner' set of data, which can then be used for analysis.

Traditional statistical wisdom prefers to analyze nuisance effects together with the effects of interest. For example, Kerr et al. (2000) define the following linear model,

$$\log(y_{ijkg}) = \mu + A_i + D_j + V_k + G_g + (AG)_{ig} + (VG)_{kg} + \epsilon_{ijkg},$$

both to evaluate differential expression, $(VG)_{kg}$, as well as to measure a score of nuisance effects, such as the array effect, A_i, dye effect, D_j, and spot effect $(AG)_{ig}$. The problem with this model is that it is both too simple and too complicated. In principle, no effect can be linear, as the expressions are bounded below and above by 0 and $2^{16} - 1$ for a 16-bit image, respectively. Moreover, the efficiencies of the Cy3 and Cy5 dyes have shown a clear non-linear relationship, as can be seen for instance in Figure 3.2. The model is rather involved on the other hand. By

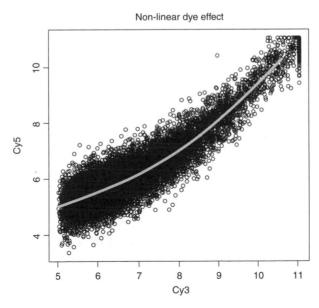

Figure 3.2 Scatter plot of the log-transformed expressions on a two-channel cDNA microarray.

including an individual spot effect, the model involves an enormous amount of parameters. Since this addition costs half of the available degrees of freedom, it might be just as easy to consider taking log ratios immediately.

Therefore, in this book, we take the approach of explicitly normalizing for the important nuisance effects *before* inference. The large amount of data available allows the estimation of non-linear dye efficiencies and spatial effects. Chapter 4 deals with this topic.

However, carry-over or so-called 'bleeding' effects of neighbouring genes are well known (Draghici et al. 2001) and this is very hard to correct by any type of normalization. Future improvements of the chip technology will undoubtedly be aimed at minimizing carry-over and other spatial effects so that the layout of the genes on the chip becomes insubstantial and does not necessarily have to be crossed.

'Dye swap' versus 'dye balance'

The design principle of blocking and crossing applies to assigning the dyes in dual-channel array experiments. 'Dye-swap' experiments measure two conditions in the two possible dye assignments. Dye swapping is an excellent way to protect against a systematic dye effect.

However, in large experiments with many conditions explicit dye swapping of each pair of conditions can be inefficient. The details of this will be discussed in Section 3.5.1. Not too much importance should be attached to the 'dye-swap principle'. As we have explained above, the dye effect can be largely eliminated with normalization techniques.

The idea of dye swapping should be replaced by 'dye balance'. In large experiments, it is impossible to perform all comparisons explicitly, let alone do a dye swap. These experiments will become more and more the norm as two-channel microarrays become cheaper. As a subset of all possible comparisons has to be selected, it is helpful if each condition is measured equally often with the Cy3 dye as with the Cy5 dye. This is called *dye balance*.

A class of designs that guarantees dye balance is the class of interwoven loop designs. An interwoven loop design is loosely defined as a set of comparisons that are lined up in one or more loops. For example, Figure 3.3 shows an interwoven loop design of 10 conditions with 2 loops. The first loop has step size 1, whereas the second loop has step size 4.

3.4   Single-channel Microarray Design

Single-channel arrays, of which the Affymetrix system is the most predominant, can be used to tackle many types of experiments. There have been implementations for simple comparative experiments, more complex multi-comparison experiments, time-course or cascading experiments, subgroup discovery experiments and experiments in the form of a diagnostic test.

Balanced dye assignment

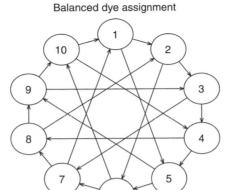

A-optimality score for contrasts: Tr[Inv(X'X)] = 22.5.
Number of arrays = 20. Number of conditions = 10

Figure 3.3   An interwoven loop design guarantees balance of the two dyes. Each arrow represents a two-channel microarray. The back of the arrow points to condition in the Cy3 channel, whereas the front points to the condition in the Cy5 channel.

3.4.1   Design issues

The design of single-channel experiments consists of three steps: (i) identifying the conditions of interest, (ii) obtaining biological replicates of each of the conditions and (iii) preparing the hybridization sample, possibly pooling biological replicates. Each of these steps involves a combination of skills and materials.

Deciding on the conditions of interest requires keen biological understanding of potentially interesting comparisons or developmental stages. In a comparative experiment, two or more interesting populations should be selected for biologically meaningful comparisons. When doing a time-course experiment, the moments of sampling are the conditions of interest and should correspond with salient biological developments. If in the future, microarrays are going to be used as a diagnostic test in hospitals, then each patient represents a separate condition of interest.

Each of the conditions in the experiment is considered a separate population from which biological replications can be sampled. The number of samples that needs to be taken depends on the amount of microarrays that are available, the number of conditions that need to be compared and the accuracy that needs to be attained. All these requirements should be balanced against one another. Pooling biological replicates into a single hybridization sample can aid inferential precision while keeping the number of chips limited, as is discussed in Section 3.2.3.

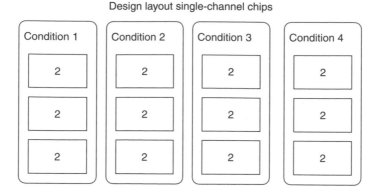

Figure 3.4   A single-channel design where four conditions are compared using three microarrays for each condition. The hybridization mixture for each chip contains a pool of two biological samples.

3.4.2   Design layout

The design of a single-channel microarray experiment can be represented by means of a graph, such as in Figure 3.4. In this hypothetical experiment, four conditions are compared with each other. Each condition is represented by three microarrays that have been hybridized with a RNA pool of two biological replicates. Smaller designs with small numbers of chips and conditions can easily be represented as such.

In the mammary gland experiment (*cf.* Section 1.2.4), the changes in expression levels over four major development stages within mice—virgin phase, pregnancy, lactation and involution of the breast—are of interest. Within each phase, several days are sampled. Three biological replicates are obtained for each of the days that have been selected. Each microarray has been hybridized with the RNA of a single replicate. No pooling was performed. The resulting 54 microarrays covering 18 time points within four developmental stages make up the experiment. Figure 3.5 represents this nested design.

3.4.3   Dealing with technical replicates

Many microarray experiments involve the use of technical replicates, that is, observations on the same RNA or on several extractions from the same biological replicate. Although the use of technical replicates at the cost of a new microarray is discouraged, several single-channel experimental systems, such as Affymetrix GeneChip, come with built-in technical replicates. These replicates can help to average out nuisance effects, such as spatial variation or spot effects.

Whether the technical replicates are on the same array or on different arrays, the best way to deal with them is by summarizing the replicates into a single value. If several technical replicates are available from a particular biological sample,

Virgin		
Week 6	Week 10	Week 12
1	1	1
1	1	1
1	1	1

Pregnancy						
Day 1	Day 2	Day 3	Day 8	Day 12	Day 14	Day 17
1	1	1	1	1	1	1
1	1	1	1	1	1	1
1	1	1	1	1	1	1

Lactation		
Day 1	Day 3	Day 7
1	1	1
1	1	1
1	1	1

Involution				
Day 1	Day 2	Day 3	Day 4	Day 20
1	1	1	1	1
1	1	1	1	1
1	1	1	1	1

Figure 3.5 The mammary gland development time-course experiment has a slightly more complicated design. The conditions *days*, are nested in the condition *developmental stage*.

then the gene expression values for that sample are best replaced by the mean or median of all the technical replicates. In this way, the number of independent replicates, or degrees of freedom, corresponds to the number of observations, and no adjustments have to be made in downstream analyses.

For the Affymetrix platform, the MAS 5.0 software stores a single summary of all 11–20 available technical replicates for each gene in a .CEL file. There are other ways to summarize the data. Li et al. (2003) describe a method that models the expression values in a multiplicative fashion from which a gene-specific expression value results. In principle, the standard errors from the group of technical replicates can be used as a quality index for subsequent analyses, as long as technical replicates are not interpreted as independent observations.

Conclusions

If each microarray is considered a blocking factor, then for one-channel oligonucleotide arrays it is not possible to apply more than one condition within each block. As a consequence, each condition should be applied to a separate array. This might mean that array effects are difficult to spot. On the other hand, a one-channel microarray experiment can be nothing but blissfully ignorant of this potential bias. For a single-channel array, there is no essential difference between comparing two or more than two conditions. Each of the conditions needs to be hybridized separately to the arrays. This leads to great simplifications in the design as compared to two-channel arrays that we shall discuss hereafter.

Several principles should, nevertheless, be kept in mind. Pooling of mRNA samples and biological replication is recommended. In order to be able to distinguish between the variation contribution of the conditions and the array effect, each condition should be replicated several times.

3.5   Two-channel Microarray Designs

The design principles we have discussed in the previous sections have focused on carefully choosing replicate experimental units, blocking structural effects, crossing these effects appropriately and avoiding any additional bias through randomization. The implications of these principles are explored in this section, where we derive practical recommendations for two-channel microarray designs. Several authors (Glonek and Solomon 2004; Kerr and Churchill 2001a,c; Yang and Speed 2002) focus on two-channel microarrays and come up with several practically useful recommendations. Our discussion will extend some of their recommendations and provide a practical tool to derive actual designs.

3.5.1   Optimal design of dual-channel arrays

Optimal design is an established area of statistical research. The term refers to the optimal assignment of the units, here microarrays, to the treatments or conditions in such a way as to optimize a particular score function. A-optimality, for example, considers the sum of all the variances of the parameter estimates of interest. This form of optimality has been considered before by Kerr et al. (2000) in a microarray setting. We shall also consider D-optimality, which aims to minimize the determinant of the matrix $(A^t A)^{-1}$, where A is the design matrix. This is not extremely different from A-optimality, although it also punishes for estimates that are highly correlated.

A dual-channel array will yield two intensity values for every spot. The two values are associated with the amount of mRNA that is produced for each gene under either condition. It has been shown in Section 6.1.1 that if one would like to minimize the risk of bias as a result of spot unevenness, then it is safer to consider the ratio of the two spot intensities as a single observation, rather than as two separate observations. However, ratios have problematic behaviour as the denominator gets close to zero and treat the numerator and denominator in an asymmetric way. For this reason, we consider the log transformation of ratios.

We assume, for simplicity, that the variation of a spot log ratio is the same across different arrays for a particular gene, say σ_g. For the moment, we focus our discussion on the expression of a single gene and, therefore, drop the reference to the specific gene g from our notation. The distribution of the log ratio can come from a wide family of symmetric distributions with a finite second moment, but without loss of generality we assume that it is normal.

Let x_i be the log-expression ratio, $\log(y_i^r / y_i^g)$, of a particular gene on array i which has been assigned to treatment $r(i)$ in the red channel and to treatment $g(i)$

in the green channel. We model the log ratio x_i as

$$x_i \sim N(d_{r(i),g(i)}, \sigma).$$

The parameter $d_{r(i),g(i)}$ represents the average log difference in expression between treatments $r(i)$ and $g(i)$. In particular, we can write

$$d_{r(i),g(i)} = \theta_{r(i)} - \theta_{g(i)},$$

where θ_j represents the mean log expression of the selected gene under condition j. This way of defining d_{ij} makes it explicit that a form of transitivity holds, namely $d_{ij} = d_{ik} - d_{kj}$.

Unidentifiability

If we consider c different conditions, then the parameters of interest are

$$\theta_1, \theta_2, \ldots, \theta_c.$$

Unfortunately, it is impossible to estimate all these variables if we only observe log differences. To see that the variables are unidentifiable, consider the following simple example. If we observe a log-difference of 5 between conditions 1 and 2 and a log-difference of -3 between conditions 2 and 3, then this could mean that $\theta_1 = 6, \theta_2 = 1$ and $\theta_3 = 4$, *or* that $\theta_1 = 9, \theta_2 = 4$ and $\theta_3 = 7$ *or* an infinite number of other possibilities.

We can impose, for example, the constraint that $\theta_1 = 0$. This is identical to using the parameters $d_{21} \equiv \theta_2, d_{31} \equiv \theta_3, \ldots, d_{c1} \equiv \theta_c$ for the parametrization.

Matrix formulation of a two-channel microarray experiment

If n dual-channel arrays are available, then we can observe n log ratios, $x_1, \ldots, x_n$. Considering the parameter constraints from the previous section, we model the log ratios as

$$\begin{aligned} x_i &= d_{r(i),g(i)} + \epsilon_i, \\ &= d_{r(i),1} - d_{g(i),1} + \epsilon_i, \quad i = 1, \ldots, n, \end{aligned} \quad (3.10)$$

where $\epsilon_i \sim N(0, \sigma^2)$ and $d_{11} = 0$. Expression (3.10) can be written in matrix form, that is,

$$x = Ad + \epsilon, \quad (3.11)$$

where $d = (d_{21}, d_{31}, \ldots, d_{c1})$ and the matrix A is called the *design matrix*. For example, the three gene expression ratios in the loop design experiment in Figure 3.6 can be represented by means of the following matrix operation:

$$\begin{pmatrix} x_1 \\ x_2 \\ x_3 \end{pmatrix} = \begin{pmatrix} -1 & 0 \\ 1 & -1 \\ 0 & 1 \end{pmatrix} \begin{pmatrix} d_{21} \\ d_{31} \end{pmatrix} + \epsilon.$$

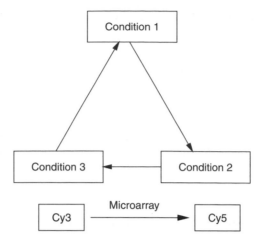

Figure 3.6   A three slide loop design.

Several people have proposed such a description of a microarray experiment (e.g. Bretz et al. 2003). Some have considered a more complicated design matrix, including, for example, dye effects and other nuisance parameters. We explicitly do not consider those parameters as we believe that in such experiments these are best dealt with *off-line*. In the chapter about normalization, we deal explicitly with ways on how to deal with nuisance parameters.

The advantage of this way of writing the microarray experiment, namely in terms of a design matrix, is that it allows one to use standard statistical techniques to evaluate the *efficiency* of the design.

Optimal design principles

One of the primary technical aims of a microarray experiment is to get adequate estimates of the gene expressions $\theta_1, \theta_2, \ldots, \theta_c$ across all the c conditions and for each of the genes. In other words, one would like to obtain the most precise estimates for all the $\binom{c}{2}$ mean log ratios,

$$
\begin{array}{cccc}
d_{21} & d_{31} & \ldots & d_{c1} \\
       & d_{23} & \ldots & d_{c2} \\
       &   & \ddots & \vdots \\
       &   &   & d_{c,c-1}
\end{array}
$$

For defining the model we only need the top row, $d = (d_{21}, \ldots, d_{c1})$, as all the other parameters are linear combinations of d. Standard statistical theory of linear models (e.g. Weisberg 1985) shows that the best estimate $\hat{d}$ of $d = (d_{21}, \ldots, d_{c1})$ is given as

$$
\hat{d} = (A^t A)^{-1} A^t x.
$$

The estimate $\hat{d}$ is a linear combination of the normally distributed x, and therefore it can be shown that $\hat{d}$ is normally distributed with the correct mean and a variance that depends crucially on the design matrix A, namely

$$\hat{d} \sim N(d, (A^t A)^{-1}\sigma^2).$$

For example, in the three slide loop design from Figure 3.6 the estimate of the parameters $d = (d_{21}, d_{31})$ under the assumption of normally distributed log-ratio expressions is given as

$$\begin{pmatrix} \hat{d}_{21} \\ \hat{d}_{31} \end{pmatrix} \sim N\left(\begin{pmatrix} d_{21} \\ d_{31} \end{pmatrix}, \sigma^2 \begin{pmatrix} 2/3 & 1/3 \\ 1/3 & 2/3 \end{pmatrix} \right).$$

Design optimality relates to 'minimizing' the variance component, $(A^t A)^{-1}$. This quantity is multidimensional and should therefore be summarized into a one-dimensional score. Two commonly used summaries are the so-called *D-optimality* and *A-optimality*.

Definition 3.1 *A D-optimal design is a design whose design matrix minimizes the determinant of $(A^t A)^{-1}$, or*

$$A_{D\text{-}opt} = \arg\max |A^t A|.$$

Definition 3.2 *An A-optimal design is a design whose design matrix minimizes the sum of all variances of the estimates of the parameters,*

$$A_{A\text{-}opt} = \arg\min Tr\left((A^t A)^{-1} \right). \tag{3.12}$$

Sometimes there are more parameters of interest than are needed for the parameterization of the model, that is, $d = (d_{21}, \ldots, d_{c1})$. For example, if we are interested in all possible contrasts, one could wish to minimize the sum of all the variances of the estimates of these contrasts. By choosing an appropriate matrix C, the vector Cd can represent any combination of the contrasts of interest. An *A-optimal design* for Cd is defined as

$$A_{A\text{-}opt} = \arg\min \text{Tr}\left(C(A^t A)^{-1}C^t \right). \tag{3.13}$$

The score in both definitions is maximized over the permissable set of designs. This set depends typically on the budget as well as on the available RNA. In what follows, we assume that the choice of a particular design depends only on the number of available arrays.

The parameters of interest in the three-array loop design are (d_{21}, d_{31}, d_{32}). Therefore, the matrix C in Equation (3.13) is given as

$$\begin{pmatrix} d_{21} \\ d_{31} \\ d_{32} \end{pmatrix} = \begin{pmatrix} 1 & 0 \\ -1 & 1 \\ 0 & 1 \end{pmatrix} \begin{pmatrix} d_{21} \\ d_{31} \end{pmatrix}. \tag{3.14}$$

The variance–covariance matrix of the parameters (d_{21}, d_{31}, d_{32}) is proportional to

$$C(A^t A)^{-1} C^t = \begin{pmatrix} 2/3 & 1/3 & 1/3 \\ 1/3 & 2/3 & 1/2 \\ 1/3 & 1/3 & 2/3 \end{pmatrix}.$$

Therefore the sum of variances in this design is 2, which corresponds to the value of the 'average variance' of 0.67 in Table 1 in Yang and Speed (2002). In this three-array loop design, the determinant score is

$$|A^t A| = \begin{vmatrix} 2 & -1 \\ -1 & 2 \end{vmatrix} = 3.$$

For a D-optimal design, it is not necessary to select the parameters of interest. By taking the determinant, it also takes the correlations between parameters into account.

Finding optimal designs

Finding optimal designs is a non-trivial task, particularly for designs with many microarrays and many conditions. Kerr and Churchill (2001a) show that it is possible to search for A-optimal design exhaustively only when the number of slides and conditions are limited, that is, typically less than 10. This is not particularly realistic for most microarray designs. Others, such as Bagchi and Cheng (1993), have suggested heuristics to find efficient, but not necessarily optimal designs.

We follow an approach suggested in Wit et al. (2004a) to find designs that are close to optimal. They use a technique, called *simulated annealing* (Kirkpatrick et al. 1983), to search the enormous design space of all possible ways to assign c conditions to a certain number of arrays. This space contains many local optima and the techniques aims to find the optimal design. Simulated annealing is an iterative search algorithm that progressively makes stronger requirements for accepting a new design. The acceptance probability for a new design is given as

$$\text{Acceptance probability} = \min \left\{ 1, \left(\frac{f(A_{\text{new}})}{f(A_{\text{old}})} \right)^{1/T} \right\},$$

where T is a tuning parameter, called the *temperature*, which is slowly lowered down to zero. The function f is the objective function that needs to be maximized, that is,

$$f(A) = |A^t A| \qquad \text{D-optimal designs}$$

$$f(A) = \frac{1}{\text{Tr}((A^t A)^{-1})} \qquad \text{A-optimal designs.}$$

Initially, almost every proposed move is accepted, whereas near the end only moves that improve the score are considered. The technique is fast and can be tuned to any required precision.

Examples of optimal designs

Reference designs have been the preferred design in the early days of microarray experiments. Nevertheless, loop designs are far more efficient in estimating the parameters of interest when comparing a small to moderate number of conditions. This is true for any optimality criterion that is considered. For example, Figure 3.6 shows the A-optimal and D-optimal design in the three-array–three-condition case.

The problem of finding an optimal design becomes more pertinent when (i) there are more conditions and (ii) there are more slides. Kerr and Churchill (2001a) already noticed, for example, that a loop design stops being optimal when there are more than eight conditions. Their conclusion was based on an exhaustive search of all possible designs. Unfortunately, this becomes completely intractable for large experiments.

Finding optimal designs with simulated annealing (Wit et al. 2004a) can overcome the computational issue. Figure 3.7 shows the optimal design found by simulated annealing for an experiment with 15 conditions and 45 microarrays. The design represented may not actually be the real optimum, but several different starting points guarantee that it must be close. Although this may not be clear immediately from the image, the optimal design assigns to each condition exactly six channels. This means that the optimal design is a form of an interwoven loop design.

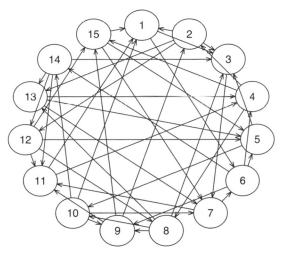

Stochastic search of A-optimal design for contrasts with simulated annealing. Acceptance rate = 43.6%. Score = Tr[Inv(X' X)] = 35.1455. Number of iterations = 1e+06. Number of arrays = 45. Number of conditions = 15

Figure 3.7 Finding A-optimal designs become more intractable for large experiments. Simulated annealing is able to find a close-to-optimal solution for an experiment with 45 arrays and 15 different conditions.

Calculating a (close to) optimal solution, such as the one in Figure 3.7, might be of limited use only. It can be difficult to convince biologists to use exactly this design, and rightly so. The likelihood of making an administrative error in performing such an intricate experimental design can be quite large. Nevertheless, an optimal design can be used as a type of benchmark. One can compare the score of an intuitive design to the score of the optimal design and then decide whether a lot is lost by using the former.

Implementation of optimal design in R: the od function

Simulated annealing is a powerful optimization tool. Kirkpatrick et al. (1983) derive conditions under which it is guaranteed that the global maximum would be found, even in a space full of local maxima. Wit et al. (2004a) implement a tool in R for finding A- and D-optimal designs for two-channel microarrays. The function od calculates an (approximate) optimal design for nt conditions and ns slides. By default, it uses n.iter = 1000 iterations, which is sufficient for small designs. However, for larger designs a larger number of iterations should be specified, as well as a value for ok.final.temp, which is closer to zero.

3.5.2   Several practical two-channel designs

In this section, we consider practical microarray designs. Rather than proposing only optimal designs, we also consider designs that are marginally suboptimal, but with other desirable properties, such as simplicity or biological relevance.

Comparing two conditions

Many microarray experiments are performed to answer relatively simple scientific questions. Often it is of interest to compare the gene expression profile for a treatment to some control condition. Two-channel cDNA microarrays are perfectly suited for this kind of experiment. On each array, both target conditions can be applied simultaneously, which makes controlling for the array block effect trivial. Pooled biological replications reduce the overall variability. And by swapping dyes evenly across the replicates, the variation introduced by the dyes is efficiently controlled. This design is known in the statistical literature as a complete block design, since within each array it is possible to measure all the conditions of interest.

The skin cancer experiment is a good example of a simple but effective two-condition microarray experiment. It uses four microarrays. On each of the arrays, both the conditions, that is, cancer and normal, are applied. Moreover, each condition is measured twice in the Cy3 channel and twice in the Cy5 channel.

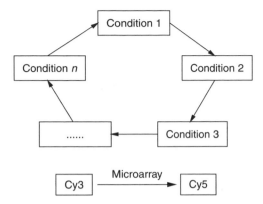

Figure 3.8 A general loop design. Each of the arrows stands for a single two-channel array, where the origin indicates the Cy3 dye.

Comparing more than two conditions

When more than two conditions are considered, it means automatically that no direct comparisons of all conditions can take place within a single two-channel chip. Any design of this type is called an incomplete block design. Initially, this problem was solved by considering a fixed reference sample, to which each condition could be compared. This design is known as the *reference design*. Kerr and Churchill (2001a) quickly pointed out that such a design can be quite inefficient, because it puts a lot of resources into measuring a reference sample, which is essentially of no interest. Instead, by dropping the reference sample altogether and comparing each sample with one of the others, a much more efficient design can be achieved. A special case of such design, shown in Figure 3.8, has been named *a loop design* by Kerr and Churchill, because it lines up the conditions in a loop. The design uses the same amount of microarrays as there are conditions.

For a moderate number of conditions, the loop design results in the least amount of variance of gene-specific differential expression estimates. It is therefore an A-optimal design. Moreover, in a loop design each condition is evenly crossed with each of the dyes. Such balance increases the overall accuracy and ease of calculation. However, if the number of conditions is more than eight, then a loop design is no longer A-optimal. For experiments with nine or more conditions, a form of a reference design, such as in Figure 3.9, is A-optimal.

When there are too many conditions, the efficiency of loop designs decreases. When one wants to compare conditions on either end of the loop, then one has to go through many intermediate comparisons in which error accumulates. If more resources are available, several loops can be considered (e.g. Figure 3.10). A so-called interwoven loop design can remedy some of the problems that are caused by long distances between the conditions in the loop. The interwoven loop design effectively replicates the loop design for a different ordering of the conditions (*cf.*

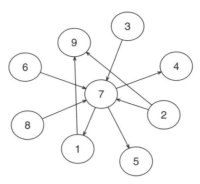

Stochastic search of A-optimal design for contrasts with simulated annealing.
Acceptance rate = 24.7%. Score = Tr[Inv(X' X)] = 57.5. Number of iterations = 5,000.
Number of arrays = 9. Number of conditions = 9

Figure 3.9   With nine conditions and nine arrays, a loop design is not A-optimal anymore.

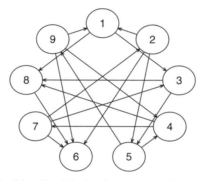

Stochastic search of A-optimal design for contrasts with simulated annealing.
Acceptance rate = 35%. Score = Tr[Inv(X' X)] = 17.6757. Number of iterations = 5,000.
Number of arrays = 18. Number of conditions = 9

Figure 3.10   If two or more arrays are available per condition, a variation of a loop design, the interwoven loop design, is A-optimal.

Figure 3.11). Although optimality properties of such interwoven loop designs are harder to prove, it can be balanced with respect to the dye factor.

Designs for time-course experiments

Time-course experiments also fall within the remit of the previous section. Each of the time points can be seen as a different condition. Yang and Speed (2002) suggest alternative designs, but recommend the loop design for time-course experiments. Loop designs for time-course experiments have the same desirable properties as for any multi-condition experiment.

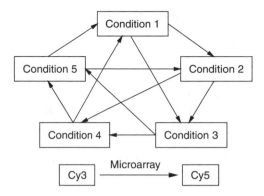

Figure 3.11   Typical example of an interwoven loop design. Each of the arrays stands for a single two-channel array, where the origin indicates the Cy3 dye.

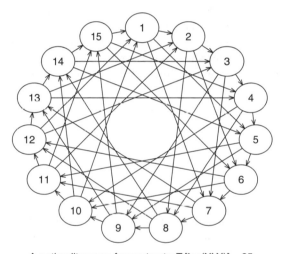

A-optimality score for contrasts: Tr[Inv(X' X)] = 35.
Number of arrays = 45. Number of conditions = 15

Figure 3.12   Design for an experiment with 15 time points and 45 slides that has the highest A-optimality score in the class of interwoven loop designs.

Large experiments complicate the choice of design. Explicit searches of the design space become impossible and approximately optimal designs can be rather daunting (e.g. Figure 3.7). For this reason, Khanin and Wit (2004) suggest reducing the design space for large time-course microarray experiments to the class of interwoven loop designs. From empirical evidence, it is clear that the optimal interwoven loop design is generally very close to the true optimal design. Figure 3.12 represents the best interwoven design. Its score, $\text{Tr}(A^t A)^{-1} = 35$, compares favourably with the computationally intensive and biologically unintuitive 'optimal' design with a score of 35.15.

In time-course experiments, not all comparisons may be equally important. Since the conditions are naturally ordered, the biologist could weigh neighbouring time points and time points that are further apart differently. For example, if the general pattern is important, it might be that the biologist is more interested in changes from the beginning to the end. However, if the biologist would like to identify where a sudden change in expression occurred, then she might be more interested in the comparisons of neighbouring time points.

We propose to look at A-optimal designs where each variance of the parameter d_{ij}, that is, the mean difference between the log expression at time point i and time point j, is weighted by

$$w_{ij} = w^{|i-j|} \tag{3.15}$$

Values of $w > 1$ greater than one correspond to putting more weight on comparisons between time points that are further apart, whereas values of $0 < w < 1$ less than one stress comparisons of neighbouring time points.

We can describe a weighted A-optimal design with the help of matrix algebra. In Section 3.5.1 we have introduced matrix notation for a microarray experiment. A is the design matrix of the experiment if the vector of observed log-expression ratios, $x = \log \text{Cy3/Cy5}$ can be written as

$$x = Ad + \epsilon,$$

where d is the average log-expression difference between all time points and time point 1. The variance–covariance matrix of the estimate $\hat{d}$ of d is given as

$$V(\hat{d}) = \sigma^2 (A^t A)^{-1}.$$

Let C be the matrix that defines all contrasts, such as, for example, in Equation (3.14). The variance–covariance matrix of all contrasts is proportional to

$$C(A^t A)^{-1} C^t. \tag{3.16}$$

An *un-weighted* A-optimal design, minimizes the trace of the matrix in Equation (3.16). Let W be a diagonal matrix whose diagonal elements are selected according to weights in Equation (3.15). A *weighted A-optimal design* associated with weights W is defined as the matrix

$$A_{\text{A-opt(W)}} = \arg \min \text{Tr}(WC(A^t A)^{-1} C^t)$$

$$= \arg \min \text{Tr}(C^t WC(A^t A)^{-1}). \tag{3.17}$$

The matrix algebraic equality in Equation (3.17) can lead to substantial computational savings, as the matrix $C^t WC(A^t A)^{-1}$ is typically smaller and the matrix $C^t WC$ can be calculated once off-line.

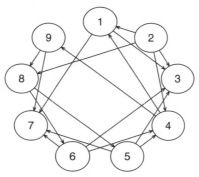

Stochastic search of A-optimal design for weighted contrasts with simulated annealing.
Acceptance rate = 35%. Score = Tr[W Inv(X'X)] = 17.1975. Number of iterations = 10,000.
Number of arrays = 18. Number of conditions = 9

Figure 3.13 Weighted A-optimal design for an experiment with 9 time points and 18 slides arrays.

Optimal interwoven loop design in R: the od function. The function od can also handle the search of A- and D-optimal designs in the class of interwoven loop designs. It requires the specification of the number of conditions, nt, and the number of arrays, ns, from which the number of loops are calculated, ns/nt. Also, the restriction to a search of an optimal design in the class of interwoven loop designs should be specified as method = "loop".

Weighted A-optimal designs in R: the od function. By specifying the parameter weight, the function od searches for weighted A-optimal designs. A positive number will act as the constant w in Equation (3.15). A value less than one will favour nearby comparisons, whereas a value greater than one will favour long-distance comparisons. Figure 3.13 shows a weighted A-optimal design for a value of $w = 0.75$. The design is quite complicated and it is possible to use instead a more intuitive interwoven loop design. Figure 3.14 of an interwoven loop design with steps 1 and 3 is the result of od(9, 18, method = "loop", weight = 0.75). Computationally, finding the best interwoven loop design is much faster, and the resulting score of 17.47 is not much worse than 17.20, achieved by the A-optimal design in Figure 3.13.

Using microarrays as diagnostic test

When the price of microarrays drops and the technology matures, it is possible that they become standard diagnostic tools in the medical profession. Microarrays have already been applied successfully in predicting cancer classes on the basis of individual expression profiles (Golub et al. 1999). When performing diagnostic

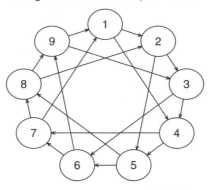

A-optimality score for weighted contrasts: Tr[W Inv(X'X)] = 17.4681.
Number of arrays = 18. Number of conditions = 9

Figure 3.14   Weighted interwoven loop design for an experiment with 9 time points and 18 slides arrays that is close to being A-optimal.

tests, it is often impractical and generally undesirable to wait for a group of patients for whom the expression profiles can be generated. In order to make the diagnostic test flexible, it should be almost instantly available to the medical practitioner. If two-channel slides are going to be used for this purpose, then this might be the only place where a pure reference design has still a role to play. If an unlimited supply of high quality reference mRNA or genomic DNA is available, then the patient's sample can be hybridized against this reference to generate a profile that is instantaneously comparable with those from other patients. If a one-channel slide is used for these purposes, then no consideration has to be given to the reference sample.

4

Normalization

Microarray experiments aim to measure the amount of transcribed mRNA. As a crucially optical technique, it actually measures the colour intensity given off by the chip on which the mRNA is hybridized. Many intermediate steps are needed in order to associate gene expression with the emission of a certain colour. In fact, in each of these steps random and sometimes systematic errors accumulate. Random errors tend to increase the variation of the results, but do not really affect their overall mean accuracy. Systematic bias, however, can spell disaster, because they can alter the structure of the results to the point that the raw data can say very little about the actual amount of transcribed mRNA.

In this chapter, we discuss normalization techniques to deal with systematic artifacts on the array. Normalization is typically performed on the spot statistics, after image analysis has transformed the pixel values into these summaries. However, some artifact removal can be done more efficiently on the pixel values themselves *during* image analysis.

4.1 Image Analysis

The task of image analysis is to convert the enormous number of pixels in the microarray images into expression values for each gene or EST. Typically, microarray image analysis programmes give a few summary statistics of the pixel intensities for each spot and for the surrounding background. In this section, we discuss several sensible ways in which image analysis can be performed. Further, the pixel intensities also depend on the actual scanning equipment used. For a good overview of scanning methods and more on image analysis, please consult Schena (2003).

Generally, there are four stages in image analysis. *Filtering* denotes a low-level image 'cleaning' procedure. Filtering can be used to remove very small contamination artifacts, such as specks of dust. It is also a term used to describe

Statistics for Microarrays: Design, Analysis and Inference   E. Wit and J. McClure
© 2004 John Wiley & Sons, Ltd   ISBN: 0-470-84993-2

robust ways of removing background trend. After filtering, the location of the centre of each spot on the array is identified; this is known as *gridding*. After this, pixels from the local area surrounding each spot are chosen to be elements of the spot or background. This is known as *segmentation*. Finally, *quantification* summarizes the pixels' intensities for each spot and its associated background as a single quantity.

Many of the image analysis algorithms are implemented in proprietary software. One of the problems with discussing proprietary software is that the algorithms are seldom given in detail. It is therefore difficult to give any advice on which software to use.

4.1.1   Filtering

Filtering is the replacement of each pixel in an image with a value derived from the pixel and other pixels surrounding it. It makes images smoother and removes local noise or interference in an image. Two types of filter are very useful for microarray image analysis—a *median filter* and a *top-hat filter*. The former allows for the removal of contamination that affects only a very small number of pixels. The latter allows for the robust removal of background on an array.

Median filter

A median filter simply replaces each pixel with the median value among the pixels in a square centred on the pixel in question. By making the square small (3×3 or 5×5, say), small artifacts on the array can be 'smoothed out' so that they disappear or are less noticeable. Since a hybridized spot with typically 200 pixels is much larger than the smoothing window, the filter will not affect the overall structure of the spot.

The dampening that a median filter provides reduces the variation in the image, which can be useful, especially if trying to estimate intensities robustly. Glasbey (2001) considers its use in more detail. Care should be taken when using the variation in spot intensity as a quality measure in normalization or inference.

Top-hat filter

The top-hat filter estimates the trend via *morphological opening* and then removes this trend from the image (Glasbey 2001; Yang et al. 2000). Morphological opening works in two stages. First, it replaces each pixel by the *minimum* value of the pixels in a square centred around it. The thereby-created image is itself transformed, this time by replacing each pixel with the *maximum* value in the window. By making the sides of the window greater than the diameter of the spots, only the trend in the data is estimated; all spots will 'disappear' in a morphological opening image. The top-hat filter then subtracts the morphological opening from the original image.

To illustrate the top-hat filter, Figure 4.1 shows a top-hat filtering of pixel values in one dimension. The solid grey line represents the original pixel values.

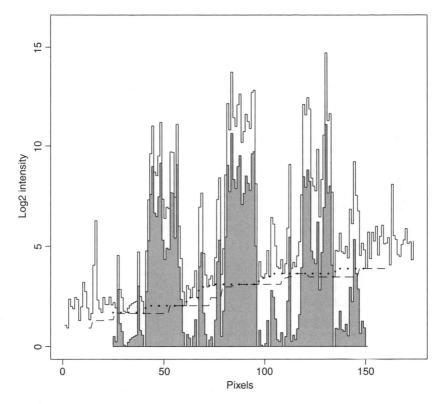

Figure 4.1 An example of one-dimensional top-hat filtering. The solid grey line represents the raw pixel values; the dashed black line is the minimum filter; the dotted black line is the subsequent maximum filter; the grey-filled line at the bottom represents the top-hat filtered values.

The dashed black line is the minimum filter, and the dotted black line is the subsequent maximum filtering of these minimum values. The grey-filled shape at the bottom shows the top-hat filtered values, where the dotted black line has been subtracted from the original grey line.

Top-hat filtering is a form of background subtraction and has been recommended by Yang et al. (2001) and Glasbey (2001). It has the advantage of ensuring that the estimates of gene expression are non-negative and 'seems to provide a good balance in the bias–variance trade-off' (Yang et al. (2001)). However, many image analysis packages do not have this as an option and Yang et al. (2001) suggest using no background correction, at least when considering log ratios with two-channel arrays. The image analysis package Spot (Buckley 2000) does allow top-hat filtering.

Yang et al. (2000) point out that choosing the window size appropriately for top-hat filtering helps to reduce the across-array variance. If the length of the sides

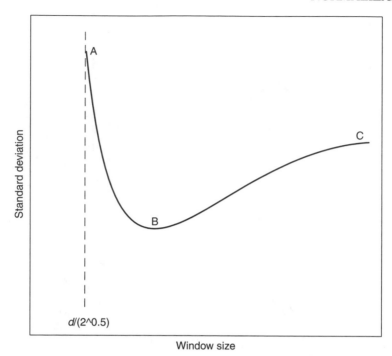

Figure 4.2   Visualization of changes in across-array standard deviation of top-hat filtering as the filter window increases beyond $d/\sqrt{2}$. At position A, the window only just captures non-spot pixels; at C, the window is so large that only a constant is removed. The optimum window size occurs at B.

is very near to $d/\sqrt{2}$, where d is the diameter of the spot, then the window will only just cover the spot. This means that very few background pixels will be sampled to choose the minimum pixel value and this will make the filter less stable and so increase the variance across the array, as is seen at position A in Figure 4.2. When the window is the size of the array (as at position C), the filtering becomes equivalent to constant background subtraction, and so the standard deviation of the array remains unchanged. At position B, the window is at its optimum size, and the filter reduces the standard deviation of spots across the array; here, the trend in the data is removed with minimal interference from local contamination.

4.1.2   Gridding

Gridding, also known as *addressing*, is determining the location of the centre of each spot on the microarray. Finding the centre of each spot is not quite as simple as moving a fixed grid over the image until it is in position. Because of the very small size of the spots, a fixed grid will typically not exactly fit the actual printing patterns. This problem can be exacerbated because the shape of a spot may not be a homogeneous circle.

Generally, a fixed grid is placed over the array, and semi-manual adjustments are made to finalize the gridding. Many packages attempt to adjust the fit of the fixed grid and allow the user to check and alter the results. Eventually, gridding partitions the array into areas, each of which contains one spot and some surrounding background pixels.

Affymetrix MAS 5.0 image analysis software does not adjust the grid shape, but only its location (Affymetrix 2001, page 119).

4.1.3   Segmentation

Having found a pixel to represent the centre of each spot and an area in which the spot is located, we need to decide which pixels in this area represent the spot and which represent the background. Four main approaches have been proposed: (i) *fixed circle*, (ii) *adaptive circle*, (iii) *adaptive shape* and (iv) *histogram* segmentation.

Fixed circle segmentation

In this case, a circle of fixed size is placed around the centre of the spot. The area inside this circle is used to calculate the intensity of the spot, and the area outside the circle is used to calculate the background associated with the spot. This form of segmentation is best used with high-quality arrays, where the size and shape of spots are highly consistent; with other arrays, the printing is rarely regular enough for this method to be useful (Yang et al. 2001).

Fixed circle segmentation is the most commonly used for cDNA image analysis packages, including ScanAlyze, GenePix and Quantarray.

Adaptive circle segmentation

This procedure differs from fixed circle segmentation by allowing the diameter of each circle to vary. This makes it more useful for dealing with lower-quality arrays, where spots may well differ in size.

Some packages, including GenePix and Dapple, allow adaptive circle segmentation, though the implementations they use differ. In Dapple's case, the diameter is based on the second derivative of the image (Buhler et al. 2000). Manual adjustments of the diameter of the circle are possible in many packages; their practicality, given the enormous number of spots, is questionable.

Adaptive shape segmentation

As the most flexible of the methods, adaptive shape segmentation allows the spot to take any shape and is usually created using a 'seed-growing' method. *Seed-growing* involves taking a starting pixel for the spot and one or more for the background surrounding the spot. The spot is then grown from its seed pixels by deciding whether adjacent pixels belong to the spot or not. The same is done for the background seed(s) until all the spots are assigned to the background or the spot.

Another adaptive shape approach uses 'watershed' segmentation to define the spot (Siddiqui et al. 2002). It is beginning to be used, though only in academically developed software so far. Angulo and Serra (2003) compare it to GenePix and ScanAlyze.

Histogram segmentation

Histogram segmentation is one of the older segmentation techniques, which has the advantage of being simpler than other methods. It plots a histogram of all the pixels in the area containing the spot and its background. Ideally, this produces a bimodal distribution of pixel values, with the higher mode corresponding to the spot and the lower mode to the background.

Unlike the other methods, histogram segmentation does not make use of the spatial nature of the pixel data. Array contamination in the background can be more difficult to identify as it can appear as part of the spot's distribution, regardless of its spatial proximity to the spot.

Quantarray has the option of using histogram segmentation, as well as fixed circle segmentation and a form of adaptive segmentation.

4.1.4   Quantification

Having decided which pixels belong to each spot and, if used, to its associated background intensity, suitable summary statistics need to be calculated. Quantification is this information extraction process.

The calculation of spot intensities is generally done using the mean of the intensities of the spot pixels. The theory behind this is that the level of fluorescence is directly proportional to the amount of hybridization for that gene and therefore also to the amount of RNA produced by the gene (Smyth et al. 2002). The calculation of each spot's local background intensity is generally done using the more robust median value, since spot contamination can distort the background values.

The standard deviation or another variability estimate of pixel intensity is a measure for the quality of that hybridization. This information can be used during inference to aid the search for truly differentially expressed genes, for example, via a weighted t-statistic (Bakewell and Wit 2004).

4.2   Introduction to Normalization

Normalization stands for the process of removing systematic bias as a result of the experimental artifacts from the data. Its history goes as far back as the history of modern statistics. In the early 1920s, scientists at agricultural experimental stations in Britain were interested in separating the differences in varieties from the unavoidable fertility differences between soil types where the varieties were planted. The analysis of variance (ANOVA) method, developed by R. A. Fisher (Fisher and Mackenzie 1923), aims at doing exactly that. It is a one-step normalization in which all the nuisance effects are considered in combination with the effects of interest in order to evaluate only the latter. It is used by Kerr, Churchill and co-workers (Kerr and Churchill 2001c; Kerr et al. 2000) in the microarray context. The main advantage of this method is that it carries over the uncertainty of the normalization into the uncertainty of the effects of interests. This avoids false confidence in the inaccurate results.

The main disadvantage of ANOVA methods is that they are computationally intensive, especially if more complicated normalizations are proposed. Moreover, for any new analysis (e.g. a new clustering, a new classification, etc.) the data will have to be normalized again, in parallel with the calculations required. Like many others, we have opted for a *sequential approach*, in which normalization of the data is done before any further analysis.

4.2.1 Scale of gene expression data

Gene expression is the process of transcription of DNA into a number of copies of mRNA. Microarrays essentially aim to measure this abundance of mRNA for each gene under the condition of interest. The observed gene expression value, however, is quite a different beast. It stands for the amount of optical 'reflection' of the dye molecules that are attached to the mRNA. Moreover, the amount of reflection is thought to be only partially linear with the actually transcribed mRNA. Distortions of linearity are particularly evident at the lower or higher intensity regions, but non-linear, dye-specific effects occur also in other intensity regions. All this leaves us with the question: 'What is the appropriate scale of gene expression data?'

We consider four possible answers to this question:

1. rank scale

2. original scale

3. scale corrected for low-level and/or high-level artifacts

4. logarithmic scale.

There are clear drawbacks to using the original intensity scale and non-parametric sceptics might argue that the only reliable information present on the array is the relative ordering of the genes. After appropriate normalizations, it is only possible to say whether a particular gene was more or less expressed than another gene on that array. However, if this is truly the case, then it would be impossible to compare genes across different arrays without additional assumptions.

Despite its drawbacks, it has been shown that the original scale is remarkably linear for a large part of its dynamic range. Depending on the specific microarray technology, linearity on the scale of 500 to 1,000 fold can be obtained (Zhu et al. 2001). If such numbers are to be believed, then intensities approximately between 50 and 50,000 are proportional to the number of transcribed mRNA molecules in the sample.

To deal with the low-end and high-end artifacts, several solutions are possible depending on the problem. Wit and McClure (2003) show that it is not uncommon for pixels in high intensity spots to exhibit saturation. This means that several or all spot pixels are measured at the highest allowable value. These authors show how survival methods can recover the uncensored signal. This method applies maximum likelihood techniques and therefore relies on several parametric assumptions. At the low end of the intensity scale, several other problems exist. Besides the non-linearity, it has been observed that the relative error increases. Although there is not much work on possible corrections of non-linearity, several ideas have been proposed with respect to the additional additive noise in the signal. Irizarry et al. (Irizarry et al. 2003a,b) discuss a probabilistic, global subtraction model for Affymetrix arrays that is also useful for two-channel arrays. This method is discussed further in Section 4.3.3. Work by Huber et al. (2002) and Rocke and Durbin (2001) deals with the same problem from a slightly different viewpoint. They propose transforming the data via a generalized log transformation in order to reduce the additive noise problem at the lower intensity. Each of these alterations of the original scale can be useful, and whether or not to apply it should be decided on a case by case basis.

However, as a general recommendation we use the logarithmic scale. The logarithmic function, like any other invertible transformations, can be done without loss of information about the original signal. The logarithmic scale is, moreover, the most natural scale to describe fold changes. Whereas on the original scale a twofold increase and a twofold decrease do not correspond with the same absolute change, this *is* the case on the log scale. Fold differences between expressions, therefore, reduce to absolute differences on the log scale.

	Original scale 50		Logarithmic scale $\log 50 = 3.9$	
2-fold decrease	25	-25	3.2	-0.7
2-fold increase	100	$+50$	4.6	$+0.7$

With the exception of background correction, all normalizations are best done on the log scale. The variance stabilizing effect of the logarithm makes the normalizations more robust to outliers. Background subtraction, however, can only be interpreted in an additive way on the original scale.

4.2.2   Using control spots for normalization

Microarrays are generally equipped with control spots. These are spots that are not of immediate interest to biologists but are put there by the slide manufacturer for one of several reasons. The so-called 'landing lights' are bright spots in the corner of the array that help in the alignment of the array while scanning.

Housekeeping genes are control spots with genetic material from non-related species that can serve as an informal quality control test for the array. However, cross-species hybridization is not uncommon, and today, it is thought that only carefully designed synthetic spots can be used for such purposes. An example is the ScoreCard developed by Amersham, which contains a set of 23 microarray controls. The controls are artificial genes that generate predetermined signal intensities that do not change across samples or experiments. Other companies such as Stratagene are also developing similar 'spiking control kits', which are sets of artificial genes together with a corresponding RNA spiking mixture that is added to the hybridization sample. The controls generate a calibration curve for determining limits of detection, linear range and data saturation. They can be used as universal references for validating and normalizing microarray data.

In the normalization procedures we describe in this chapter, we encourage the use of control spots. Although it is possible to use all genes for internal normalization, the use of control spots holds the promise of avoiding systematic bias. On the other hand, whereas control spots are very promising, they cannot perform miracles. In particular, there are several important drawbacks. Just like any other spot on the array, the control spots are subject to random noise. If the number of control spots on the array is too small, it is impossible to decide whether the fluctuations of these spots are due to systematic variation or random error. A relatively large number of control spots are needed, spread out across the array, to be able to use them for normalization.

More serious is the case when control spots themselves show systematic fluctuations. The amount of spiking control added to each sample should be constant throughout the experiment, otherwise one normalizes against a 'moving target', which might actually introduce rather than remove bias. Only highly reproducible spiking control kits can be used for this purpose.

4.2.3   Missing data

It happens frequently that some sections of the array do not produce any readings or produce faulty readings. For example, in the breast cancer study 2.7% of all the data was missing. It could be that some part of the array is damaged and that some data has to be excluded from consideration. In this section, we describe how to deal with missing and other unreliable data.

Dealing with unreliable data

Quantitative measures, such as within-spot pixel variance for cDNA arrays or within-probe variance for Affymetrix arrays, are good indicators of the reliability of a particular expression value. It is not always necessary to omit data that is more variable than usual. Techniques exist to incorporate this type of data, but 'down-weigh' its importance in the analysis. For example, Bakewell and Wit (2004) describe a method to detect differentially expressed genes that weights observations by the inverse of their spot standard deviations.

However, it is not always feasible to include observation weights into each quantitative method. In such cases, a decision has to be made whether or not to include problematic data in the analysis. A promising method is the automatic outlier and artifact detection for Affymetrix GeneChip data implemented in dChip (Li and Wong 2003; Li et al. 2003). This method compares the multiple probes for each gene across replicate arrays to detect outlier probes or outlier arrays. Outliers are defined as replicates that do not follow the overall pattern. These outliers can be automatically replaced by imputed values. Affymetrix itself has also implemented a so-called 'detection algorithm', which results in *present, marginal* or *absent* calls. The interpretation of these calls is whether or not the 'gene expression signal' in the perfect match (PM) probe set is significantly different from the 'background signal' in the mismatch (MM) probe set. An *absent call* can be generated by uncharacteristically large MM values. This could indicate that there is some physical slide problem with the probe pairs for that particular gene and that the resulting value should be removed from consideration. Another natural explanation, however, is that the MM was 'contaminated' by cDNA from the PM probe. Moreover, an absent call can be generated because of small PM values, which only indicate that a particular gene was not expressed. We recommend that, in general, absent calls are ignored, unless it is believed that in a particular application low expression genes are uninteresting. The use of MM values is discussed in Section 4.4.

Most frequently, problematic data have to be spotted by eye or by semi-formal methods, discussed in Chapter 5. For example, Figure 5.11 shows a clear artifact, probably a hair or a scratch. It is unlikely that the data coming from the scratch contain much information. Consequently, the best option is to remove the data and to create missing values.

Dealing with missing values

Several analysis methods do not suffer from missing data. Especially, methods that depend on the dissimilarity matrix of the data, rather than on the data matrix itself, can be adjusted in a straightforward manner. The (absolute) correlation dissimilarity can be calculated by ignoring the missing values. In principle, there are two choices: either one omits the variables that are missing in each pairwise comparison, or one omits the variables that have at least one missing value throughout. Any pairwise omission might lead to a non-positive definite correlation matrix. However, if only dissimilarities are needed, then this is not a problem.

For geometric distances, such as the Manhattan and Euclidean distance, a correction factor has to be included in the calculation of the dissimilarity matrix in order to account for the smaller number of dimensions in which the distance is measured. In particular, if m of the p variables are missing in either observation x or y, then the power distances between two observations should be calculated as

$$d_q^*(x, y) = \left(\frac{p}{p - m} \sum_{i=1}^{p-m} |x_i - y_i|^q \right)^{\frac{1}{q}},$$

where for $i = p - m + 1, \ldots, p$, either x_i or y_i is missing. In particular, $d_1^*(x, y)$ and $d_2^*(x, y)$ stand for the corrected Manhattan and Euclidean distance respectively. The resulting dissimilarity matrices can be used in clustering algorithms, described in Chapter 7.

Table 4.1 contains a fictitious data set with three samples and four genes. One of the values is missing. The dissimilarity matrices for the correlation measure and power distance can be calculated in a straightforward manner. Table 4.2 contains three dissimilarity matrices for the small data set, based on two correlation measures and the Manhattan distance.

Table 4.1   A small data set with a missing value.

Sample	Gene			
	A	B	C	D
1	1.00	2.00	2.00	—
2	2.00	8.00	3.00	7.00
3	1.00	3.00	2.00	5.00

Table 4.2   Several dissimilarity measures applied to the missing data Table 4.1: (a) one minus correlation dissimilarity where gene D is only omitted in the comparison of sample one and two; (b) one minus correlation dissimilarity where gene D is omitted throughout; (c) The uncorrected Manhattan distances between samples one and two and samples one and three are scaled up by a factor 4/3.

Pairwise correlation				Correlation				Manhattan		
	1	2	3		1	2	3		1	2	3
1	0.00	0.37	0.13	1	0.00	0.37	0.13	1	0.00	10.67	1.33
2	0.37	0.00	0.20	2	0.37	0.00	0.07	2	10.67	0.00	9.00
3	0.13	0.20	0.00	3	0.13	0.07	0.00	3	1.33	9.00	0.00
(a)				(b)				(c)		

Unfortunately, many methods are not easy to extend to deal with missing values. In such cases, it is important to impute reasonable values for those that are missing. A data augmentation scheme (Tanner and Wong 1987) would be able to impute data while preserving the associated uncertainty of such imputation. However, such methods are currently impractical, especially for high-throughput approaches. Several authors (Troyanskaya et al. 2001; Wall et al. 2003) developed algorithms based on singular value decomposition (SVD) for imputing missing values in gene expression data. Troyanskaya et al. (2001) implemented this method in a programme called *SVDimpute*. This method can be computationally intensive, depending on the number of samples.

K-nearest neighbours (K-NN) is a method for imputation that has a more robust performance than the SVD method (Troyanskaya et al. 2001). It has been proposed by several people (Chipman et al. 2003) and has been implemented in several standard packages (e.g. GEPAS, http://gepas.bioinfo.cnio.es). Its simplicity and good performance makes it an ideal candidate for high-throughput work. The method determines the k samples with the most similar expression pattern compared to the sample with a missing value. Then it imputes the missing value of gene g by taking a weighted mean across the expression values for gene g across the k nearest samples. The weights can be constant or set proportional to the inverse of the distance to the other samples. The latter takes into account the similarity of the expression profiles and makes the method less dependent on the choice of k. It has been suggested that a choice of k in the range of 10 to 20 gives good results (Chipman et al. 2003; Troyanskaya et al. 2001). The choice of metric to determine the k most similar samples can be one of the power distances, such as the Manhattan or the Euclidean distance. The use of the correlation dissimilarity, however, is meaningless in the K-NN method. Two highly correlated samples can have very different profiles, which results in uninformative averages. In order to minimize the effect of outliers, it is prudent to log-transform the data before imputation by K-NN.

Mathematically speaking, K-NN can be performed on the observations as well as on the variables. Specifically, when there are many variables and few observations, as is typically the case in microarray studies, this seems an attractive option. However, imputing the expression of a gene from other genes in that sample is not very intuitive from a biological point of view.

Table 4.3 shows how K-NN can be applied to the missing data in Table 4.1. The first sample is closest to the third sample and therefore 1-nearest neighbours imputes the missing value as 5. For 2-nearest neighbours a weighted average between 5 and 7 is taken, with a larger weight on the former.

Use of `impute.missing` function in R

K-nearest neighbours is implemented in the R language in the `impute.missing` function. It requires the data matrix to be supplied with samples in the rows and genes in the columns. The value k determines the number of nearest neighbours that are used in the routine. The default is 10. Either a distance object `dist.obj` can be

Table 4.3   K-nearest neighbours applied to the missing data in Table 4.1: (a) Imputed data using only one neighbour; (b) Imputed data using a weighted mean of two neighbours; (c) The resulting correlation dissimilarity using imputed data matrix b; (d) The resulting Manhattan distance matrix using imputed data matrix b.

K-NN, $k = 1$					K-NN, $k = 2$			
	A	B	C	D		A	B	C	D
1	1.00	2.00	2.00	5.00	1	1.00	2.00	2.00	5.22
2	2.00	8.00	3.00	7.00	2	2.00	8.00	3.00	7.00
3	1.00	3.00	2.00	5.00	3	1.00	3.00	2.00	5.00
	(a)					(b)			

Correlation, K-NN, $k = 2$				Manhattan, K-NN, $k = 2$			
	1	2	3		1	2	3
1	0.00	0.42	0.04	1	0.00	10.00	1.00
2	0.42	0.00	0.20	2	10.00	0.00	9.00
3	0.04	0.20	0.00	3	1.00	9.00	0.00
	(c)				(d)		

supplied or a `metric` can be specified to determine the k nearest neighbours. By setting the `method` to `knn.wt` or `knn`, the missing values are calculated according to a weighted or ordinary mean over the k nearest neighbours respectively.

4.3   Normalization for Dual-channel Arrays

Dual-channel arrays have the greatest need for normalization because of their largely non-standardized methodology and the potential for correlations between the two channels. Two-channel arrays come in many different forms and shapes, are sometimes custom-designed and are produced by many different companies and non-profit organizations, such as HGMP (Human Genome Mapping Project, http://www.hgmp.mrc.ac.uk) and BμG@S (Bacterial Microarray Group at St Georges Hospital) in Britain. A variety of arrays means a variety of quality standards, and each array type might need a different type of normalization. The ones mentioned in this chapter are the most common types needed for two-channel arrays.

It has been a long standing practice, going back to Chen et al. (1997), to consider only log-ratio data for dual-channel arrays. Although we have suggested that there is a case to be made for using the individual channel data (*cf.* Section 3.5), it is common to use ratio data. Our recommendation is to use the normalizations described in this section, before taking ratios. The reason is the existence of intensity-dependent dye effects, sometimes known as the 'banana effect'. Taking ratios before normalizing for such effect can bias the results.

Combined normalization algorithm in R: all.norm

In this chapter, we propose a collection of normalization techniques. Each of the techniques has been implemented in R and is discussed in their relevant sections. The function all.norm does a full normalization for the complete gene expression matrix of a two-channel microarray experiment. It sequentially applies all the recommended normalizations to the data.

The function requires the gene expression matrix to have arrays in the rows and genes in the columns. Each pair of rows is presumed to come from the same array, that is, a Cy3 and a Cy5 channel. The order in which the channels are presented is irrelevant. Furthermore, the location of each spot needs to be defined in terms of two vectors x and y, each of the same length as the number of spots.

If there is any information about spots whose expression levels are not expected to change across the arrays, then these spots can be specified using the invariantset argument. If no information about invariant spots is available, then it is possible to specify a number i and let the function find the i spots that have changed the least. The invariantset argument is used in dye effect normalization and in global location normalization.

The argument empty can be called to define the spots on the array that are empty. However, it is only needed for background normalization, which is not done by default.

The argument conditions is a vector of the length of the number of channels, that is, the number of rows of the expression matrix. With this vector, the user should indicate which condition is applied to each channel.

The arguments spat, bkg, dye and cond are logical variables and control whether spatial, background, dye and condition normalizations should be performed. All are done by default, except for background normalization.

4.3.1   Order for the normalizations

Typically, a microarray is subject to several artifacts, each of which can compromise the integrity of the data. It is of paramount importance that these artifacts are removed before analyzing the data. Kerr and Churchill (2001c) establish a connection with the situation in agricultural design where similar nuisance effects can be present. In agricultural design, the natural analysis tool is a factorial model combined with an ANOVA. Effectively, this method deals simultaneously with normalization, that is, modelling of the nuisance effect, and the estimation of condition-specific gene expression. Unfortunately, it has been widely recognized (Tseng et al. 2001; Wolkenhauer et al. 2002; Yang et al. 2002a) that many artifacts are non-linear or intensity dependent and that a simple linear model is not sufficient. Although it is theoretically possible to propose more complicated models in which all effects are estimated simultaneously, it is a computational nightmare.

Most current approaches (Wolkenhauer et al. 2002; Yang et al. 2002a) attempt to estimate the nuisance effects sequentially.

When the normalizations are done one after the other, it is almost inescapable that the order of the normalizations is going to affect the outcome. It is therefore crucial that some rhyme or reason is applied in selecting the order of the normalizations. We have found it useful to apply the following rule of thumb:

> *Normalize all local features first and then gradually progress to normalizations that involve several or all arrays.*

If one is not careful with normalization, it is possible that higher-order normalizations introduce bias if lower-order normalizations have not been dealt with carefully. One such case is discussed in Section 6.2.3.

In particular, we propose to normalize the data in the following order:

1. spatial (unequal hybridization) correction

2. background correction

3. dye-effect correction

4. within replicate rescaling

5. across-conditions rescaling.

Although background correction is mentioned in this list, Section 4.3.3 will urge caution and, in fact, advise against subtracting local background estimates. Many other normalization schemes have been proposed, and similarities can be found between our methods and methods of others.

4.3.2   Spatial correction

Frequently, a quick glance at a cDNA microarray image reveals patches of unequal hybridization across the microarray. Sometimes this is expected, since clusters of related genes were located near each other or because the control spots are all together in one area. At other times the varying intensities reflect more recalcitrant effects, directly or indirectly related to the optical properties of the microarray technology. Washing the chip unevenly, inserting the chip at a slight angle in the scanner, edge effects due to evaporation and other seemingly insignificant events might alter the expression values in a way that crucially depends on the location on the array. Similarly, uneven wear of the print tips after hours of printing might alter the length or the tip opening, which alters the amount of probe spotted on the array. As a consequence, these variable print-pins can cause spatial patterns.

Different approaches to spatial normalization

One of the reasons why reference designs combined with the use of ratio data have been so popular has to do with the normalization of the spatial effect. If each of the channels of the same array is affected in a multiplicative fashion by some complicated function $C(x, y)$, where (x, y) are the two-dimensional spatial coordinates of the spot, then each of the signals in the two channels, $S_d(x, y) = g_d(x, y)C(x, y)$, are distorted versions of the true expression, $g_d(x, y)$, of the gene at that position in channel d. However, the ratio of the two signals,

$$\frac{S_{Cy3}(x, y)}{S_{Cy5}(x, y)} = \frac{g_{Cy3}(x, y)}{g_{Cy5}(x, y)}$$

corresponds to the true expression ratio. Therefore, *under the assumption that spatial effects are multiplicative and affect each of the channels similarly* taking ratios is an effective way to normalize the data. Unfortunately, spatial effects tend to be more complicated than identical multiplicative effects across both channels. Figure 4.3 shows a perspective plot of the log ratios of two channels of the same array. The gradient indicates that there is still a residual spatial effect after taking ratios.

Other methods of normalization have their own advantages and disadvantages. Yang et al. (2002b) propose a print-tip normalization as a proxy for a full spatial normalization. Although this could work satisfactorily in certain cases, the danger with such a discrete approach is the introduction of bias at the edges of the print-tip area, without resolving the true spatial effect. Consider Figure 4.4. In this case,

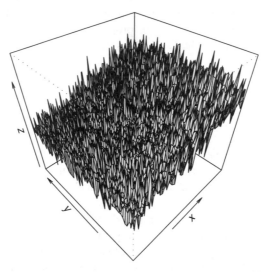

Figure 4.3   Log-expression ratios of Cy5 vs Cy3 across the third skin cancer array. It appears that the spatial effects on the array are not eliminated by taking ratios.

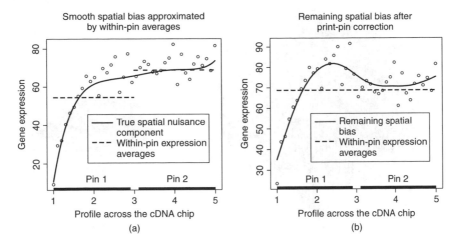

Figure 4.4 An example of an unsuccessful spatial correction by using pin information as a proxy for spatial correction.

the true underlying spatial bias was actually continuous. By subtracting an overall mean within each print-pin group, the spatial bias towards the edges of pin 1 remains, even though the overall between-pin means are the same.

Instead, robust smoothing of the expression data across the array in each of the two channels separately can deal with the spatial effect in a straightforward way. A local linear trend surface, that is, *loess* of degree one, using a significant fraction of the actual data (span = 0.5) provides a smooth spatial fit to the data, as can be seen in Figure 4.5. By subtracting the smooth surface M from the original data S,

$$S_m(x, y) = S(x, y) - M(x, y), \tag{4.1}$$

a smooth spatial trend is removed from the data. Equation (4.1) corresponds to taking the residuals of the original data with respect to the smoothed curve.

Yang et al. (2002b) suggest that spots across the array may not only differ because of a spatial location parameter but may also exhibit location-dependent scale differences, that is, regions of the array that are more variable than others. They suggest the estimation of a per print-pin scale parameter and the division of the centralized expression values on each array by their own scale parameter.

A similar idea is to smooth the absolute differences between the observed expressions and the smoothed surface. The second smoothed surface is an estimate of the location-dependent scale parameter. It has again the advantage of not introducing bias at the edge of the pin-region when the spatial scale bias is in fact smooth. By dividing the location smoothed surface S_m, as defined in Equation (4.1), by the smoothed scale surface Sc,

$$S_{ms}(x, y) = \frac{S(x, y) - M(x, y)}{Sc(x, y)},$$

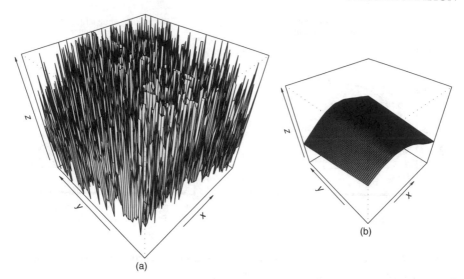

Figure 4.5   The Cy3 channel of the third array in the skin cancer study, displayed in (a), shows a spatial gradient that can be robustly estimated with a loess curve, as in (b).

the resulting surface will not have first and second-order spatial effects anymore.

Biologists might find it disconcerting that transforming the data changes their scale. Several choices are open to remedy this problem. Each value can be made positive again by multiplying by the highest smoothed scale value, max $Sc(x, y)$, and by adding back in the highest value of the smoothed surface, max $M(x, y)$. However, maxima are notoriously variable. We prefer to multiply the smoothed values by the median of the scale smoothed surface, median $Sc(x, y)$, before adding the median of the original data, median $S(x, y)$,

$$S_{\mathrm{ms}}^{\mathrm{original}}(x, y) = S_{\mathrm{ms}}(x, y) \times \mathrm{median}\ Sc(x, y) + \mathrm{median}\ S(x, y).$$

This is the default option in the R spatial normalization function, spat.norm. Although this might give negative values, this is not particularly a problem and will be dealt with later. Moreover, if the data are normalized on the log scale, negative values are perfectly fine.

Figure 4.6 shows how the spatial location and scale transformation succeed in removing the spatial bias in both the Cy3 and the Cy5 channel. In the original plot, the different layers of points correspond to different sections of the array. In the normalized plot, the data form one consistent cloud. The data do show some curvature, suggesting that there may be a dye effect. Removal of the dye effect is discussed in Section 4.3.4.

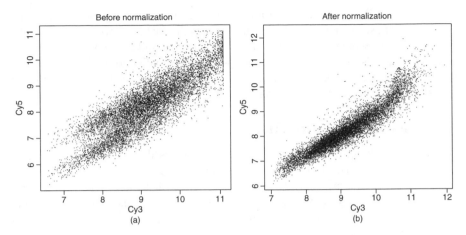

Figure 4.6   The data from the third array in the skin cancer study before (a) and after (b) spatial location and scale normalization.

The spat.norm function in R.

The function spat.norm calculates a spatial location and scale normalization (optional) in R for each channel separately. It requires the x and y coordinates for each spot value, as well as the spot intensities. After that, it will return the normalized data automatically on a scale centered around zero or on the original scale. However, more advanced users can adjust some of the smoothing parameters.

Computational details: locally weighted regression

Locally weighted polynomial regression or *loess* (Cleveland 1979; Cleveland and Devlin 1988) is a smoothing method that uses combinations of simple linear regression functions to model the deterministic part of complex response surfaces. The main attraction of this method is that the user does not have to specify the functional form of the response surface, which is attractive when one is not particularly interested in it.

Like any smoothing method, the loess depends on smoothing parameters. Effectively, there are two types of parameter in the loess case. The first is the degree of the polynomial fitted locally to the data. The loess function in R uses polynomials of the order two by default, but for the purposes of spatial normalization we recommend using linear functions, that is, polynomials of the first degree. Higher-order polynomials tend to be unstable, particularly near the edges (of the microarray) and can therefore introduce bias. The second parameter determines the fraction of the data to be included in the smoothing of each location on the array. Again, here we recommend the use of more rather than less robust values. We have chosen a value of span = 0.5 for the location smoother and a value of span = 0.75 for the scale smoother. The latter we smooth more robustly, because scale parameters are notoriously unstable.

Especially, if control spots are present in a concentrated region on the array, then, non-robust spatial smoothing methods might make incorrect adjustments. Generally, it is recommended that control spots, such as spikes, housekeeping genes or empty controls, be eliminated before applying the spatial normalization. Within the loess as well as the spat.norm functions, this can be done by specifying subset = - <control spots vector>.

4.3.3 Background correction

A common problem in measuring optical quantities is the existence of persistent background signal. This is a signal that is measured irrespective of any true signal. Early on, in microarray analysis, it was recognized that there may be a similar problem with gene expression values measured as optical intensities from incorporated dyes. It was observed that some target will attach to the array even when there is no probe available.

Several background correction methods

Several solutions have been proposed for counteracting this effect. Most solutions assume that the background effect is additive, that is, that the observed signal S is a sum of the background signal B and the true signal T,

$$S = B + T.$$

Unfortunately, the background signal *in the spot* cannot be measured. The only measurement that is available is the background value *near* the spot, B'. The simplest method (Eisen 1999; Wolkenhauer et al. 2002) simply subtracts this value from the observed value as an estimate of the true signal,

$$\hat{T} = S - B'.$$

Others suggested to subtract additionally three standard deviations of the background signal B' as a protection against a very variable background (Suite 1999). It has been observed early on that these estimates for the true signal can be negative. In some cDNA chips, as many as 10% of the genes thus get assigned a negative gene expression value. In the absence of any clear interpretation of such value, several *ad hoc* solutions have been put forward such as subtracting only a fraction of the background (Efron et al. 2001) or some overall small number. Whatever the details, these types of methods define the true expression as a combination of two quantities, each with its own associated noise. As a result, the total variation of the expression level is larger than would have been if only the foreground signal was taken.

Kooperberg et al. (2002) have applied a Bayesian analysis to this problem, which is still based on the additivity assumption of background and foreground. They effectively assume that the pixel mean observed signal value μ_S is the sum of the mean true signal μ_T plus the mean background signal μ_B and that the spot background B has the same distribution as the observed B'. The task is then to find the posterior of μ_T given S and B'.

A problem with all of these methods is that they fundamentally assume that background and foreground signal are additive. It has been recognized, however, that the DNA on the spot can effectively mask the background (Keith Vass, personal communication). If this is the case, then the background intensity *outwith* the spot might give very little indication of the amount of non-specific hybridization of the dye to the spot.

Another problem that these background correction methods do not always address is the issue of background contamination. The physical vicinity of the background measurements to the DNA probe combined with the imperfect alignment of the grid opens the possibility that some of the background signal might actually be true gene expression. A possible indication for such phenomena is positive correlation between the foreground and background signal, as shown in Figure 4.7. Special care is required to interpret these plots as some of such correlation might be due to a spatial effect and therefore may not be a sign of contamination.

Our recommendation

After these considerations, we believe that it is most prudent *not* to subtract local background values as they are prone to contamination and additional variation. Nevertheless, it is quite understandable that particularly on an array with empty spots it makes sense to assume that the lowest achievable value should be zero. Subtracting a constant amount from each gene expression value so as to make the lowest achievable value on the array zero has the additional advantage that it does not introduce additional variation. Conversely, the 'global' background normalization does not reduce the variation either.

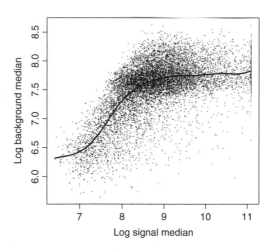

Figure 4.7   Possible indication of structural contamination of the background signal by the foreground signal.

In principle, we recommend a slight variation to this theme; that is, to subtract the mean or median of all empty spot values and to set all thus-obtained negative values to zero:

$$\hat{y}_i = \max(y_i - \mu_e, 0), \tag{4.2}$$

where $\hat{y}_i$ is the corrected spot intensity of spot i, y_i is its original spot intensity and μ_e is the average intensity of all the empty spot in this channel on the array. The main disadvantage of the deterministic method can be seen in Figure 4.8. By simply putting the lowest n numbers to zero in each channel, the noise at the lower level gets exaggerated. This might have further repercussions on the dye normalization, which takes the scatter plot as a point of departure.

Another global method that does not depend on local background values has been suggested by Irizarry et al. (2003b). The method was devised for an Affymetrix chip, but it extends to the two-channel case without alterations. It has the advantage of being motivated by a probabilistic model. However, the answers are very similar to our recommendation in Equation (4.2). The method determines the conditional expectation of the true signal given the observed signal,

$$E(s_i \mid s_i + b_i), \tag{4.3}$$

where the observed signal is thought of as the sum of the true signal s_i and the background b_i. In order to make the calculations mathematically tractable, it is assumed that the spot intensities and the background intensities on the array can be considered as drawn from one exponential distribution and one normal distribution

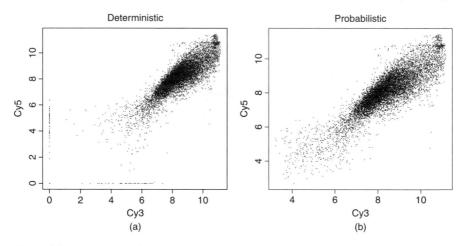

Figure 4.8 Although the simple, deterministic method (a) is an adequate method for background normalization of a single channel, the probabilistic method (b) has advantages for two-channel arrays because it avoids hard zeroes.

respectively,

$$s_i \sim \text{Ex}(\theta), \quad i = 1, \ldots, n_{\text{obs}}$$

$$b_i \sim N(\mu_e, \sigma_e^2), \quad i = 1, \ldots, n_{\text{obs}},$$

where θ, μ_e and σ_e^2 have to be estimated from the data. The spot distribution should not be confused with the pixel distribution. The parameters μ_e and σ_e can be estimated as the mean and standard deviation of the empty spots, if available. The parameter θ can be estimated as the inverse of the mean expression level of all spots minus that of the non-empty spots on the array,

$$\hat{\theta} = \frac{1}{\overline{y} - \mu_e}.$$

The background corrected intensities then correspond to the conditional expectation in Equation (4.3), which can be shown to be approximated by

$$\hat{y}_i = y_i - \mu_e - \sigma_e^2\theta + \sigma_e \frac{\varphi\left(\frac{y_i - \mu_e - \sigma_e^2\theta}{\sigma_e}\right) - \varphi\left(\frac{\mu_e + \sigma_e^2\theta}{\sigma_e}\right)}{\Phi\left(\frac{y_i - \mu_e - \sigma_e^2\theta}{\sigma_e}\right) + \Phi\left(\frac{\mu_e + \sigma_e^2\theta}{\sigma_e}\right) - 1}, \quad (4.4)$$

where φ and Φ are the probability density function and the cumulative distribution function of the standard normal distribution, respectively. The parameters μ_e, σ_e and θ can be replaced by their estimated quantities. The first two terms in Equation (4.4) dominate, and the expression therefore is, in practice, similar to Equation (4.2). In fact, preliminary simulations on simple data suggest that even if the model assumptions are fulfilled, then the deterministic method does at least as well in reconstructing the true signal as this probabilistic approximation.

The model underlying the probabilistic method shares similarities with the models proposed by Rocke and Durbin (2001) and Huber et al. (2002), although the aims of the analyses—background correction and variance stabilization respectively—are quite different. Our own experience is that the ease of implementation makes the global background correction methods in Equations (4.2) and (4.4) more attractive candidates to deal with low-intensity additive effects. Nevertheless, the field of variance stabilizing methods is very actively pursued by several people and easy, fast and stable implementations may well soon be developed. The Bioconductor package vsn (Huber et al. 2002) is an encouraging contribution.

The bkg.norm function in R.

Conforming to our general recommendation about background normalization, we have implemented a global background normalization function, bkg.norm in R. It can perform both recommended normalizations—a deterministic normalization as in Equation (4.2) or a probabilistic approach, as described by Equation (4.4).

If no information is provided about empty spots, then it subtracts on the original scale a constant from each expression value such that the lowest expression value is some preset minimum, typically zero or one. For the deterministic method, if a set of empty spots is specified in empty together with a certain quantile q (default $q = 0.50$), then the qth quantile of the empty spot intensities is subtracted from the data, and all the negative values are set to zero. The probabilistic method uses the set of empty spots to estimate μ_e, σ_e and θ and then normalizes the spot values y_i according to Equation (4.4).

If for the parameter empty, only a number n is specified, then it is assumed that the n lowest expression values correspond to the empty spots and, as before, the deterministic or probabilistic method can be applied.

Any type of background correction discussed in this section should be done on the original scale and *not* on the log scale. Subtraction on the log scale corresponds to division by a constant, which is inappropriate for the high intensity spot values. If data are supplied to the function bkg.norm on the log scale, then the parameter dat.log.scale should be set to true.

4.3.4 Dye effect normalization

The current technology of dual-channel microarrays is based on measuring optical intensities of dye labelled cDNA that has hybridized to gene-specific probes on the microarray. The two most commonly used dyes in dual-channel arrays are the Cy3 and Cy5 dyes. Despite similarities, the dyes have slightly different properties. The quantum yield from the dyes is different, the size of the Cy3 and Cy5 molecules differs slightly, which leads to differential incorporation of the dye molecules on the array. Also, the dyes react slightly differently to photo-bleaching, an effect that occurs as a result of multiple scannings of the array. As a consequence, the two channels of a Cy3–Cy5 array have slightly different efficiencies, which makes direct comparison of the gene expression data difficult, if not impossible.

Dye normalization methods

Several methods have been proposed to deal with this dye effect. Early papers (Kerr and Churchill 2001a) suggested correcting the effect by a constant, possibly different for each array ('dye-array interaction'). This means that the dyes are assumed to have efficiencies that differ by an additive or multiplicative constant. This corresponds to a linear relationship between the, possibly log-transformed, expressions in the Cy3 and Cy5 channels. Figure 4.9(a) shows a case for which this relationship approximately holds. However, plot (b) in the same figure shows a clear deviation from a constant dye effect.

Global normalization methods are not adequate when the dye bias depends on the overall spot intensity. Figure 4.9(b) shows the very common case in which the relative efficiency of the dyes seem to vary across the intensity range. Whereas, in this case, overall the Cy3 dye seems to have been incorporated more efficiently, the Cy5 dye seems to have gained in efficiency relative to Cy3 in the middle of the intensity range. This kind of effect is known in bioinformatic circles as the 'banana effect', for obvious reasons.

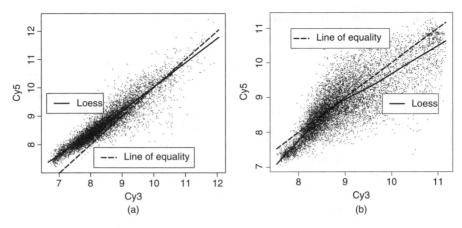

Figure 4.9   Raw log-transformed data from two different cDNA slides from the same skin cancer experiment. The x-axes and y-axes contain the Cy3 and Cy5 values respectively. The lines in both plots correspond with the line of equality and a loess smoother through the points. Plot (a) of array 2 shows a remarkably linear dye effect, whereas plot (b) of array 1 is clearly indicative of a more complex, non-linear relationship between the dyes.

Two methods have been suggested to deal with intensity-dependent dye effects. The first method consists of estimating the relative dye efficiency at each intensity and subtracting it from the data. A variation of this method is discussed in the following section. The other method is sometimes called *dye-swap normalization* (Yang and Speed 2002).

Dye-swap experiments consist in repeating a hybridization twice with the dyes swapped and averaging the expression values for each spot over the Cy3 and Cy5 channel. There are two main problems with this method. First, the dye effect tends to differ from array to array, and there is no guarantee that the method effectively removes the dye effect. Secondly, a dye-swap experiment is not the most efficient way of measuring differential expression in large designs, as was explained in Section 3.3.2. Nevertheless, if a dye-swap experiment has been performed, then averaging out expressions in the Cy3 and Cy5 channels gives some protection against under- or over-smoothing.

Two words of caution are appropriate here. Whereas this intensity-dependent dye bias is well known, it might be that there are still some other effects that influence the incorporation of dye molecules onto the array. It would be essential to normalize also for these effects in order to avoid bias. At the same time, the dye bias can be confounded by other nuisance effects. In Section 6.2.3, we saw a case in which the dye effect was confounded with a spatial effect on the array. In such instances, direct application of dye normalization methods can be disastrous and, in fact, introduce bias.

Recommended intensity-dependent dye normalization

The intuitive form of intensity-dependent normalization by fitting a smooth curve to a scatter plot of Cy5 versus Cy3 values, such as in Figure 4.9, has several disadvantages. The method is not invariant under the exchange of the axes. Given that neither Cy3 nor Cy5 is a natural response value, this is not very satisfactory. A related issue is the fact that the usual residuals are not the smallest distances to the smoothed line. Orthogonal distances could be used, but this tends not to be standardly implemented. Instead, Yang and Speed (2003) suggest the smoothing of the data on a transformed scale, sometimes known as the MA scatter plot.

The intensity-dependent dye normalization uses all or part of the data first to estimate a line of equal expression and then to define individual gene expressions as deviations from that line. The user should decide which data to use for the normalization. There are two criteria for the choice of a normalization set: (i) the expression of the genes should be expected to be approximately equal across both dyes (ii) the normalization set should not be too small. Whereas the former requirement minimizes the risk of 'normalizing away' true differential expressions, the latter prevents experimental noise, having a significant impact on the normalization curve.

For each probe i in the invariance set N, we transform the raw Cy3 and Cy5 values, G_i and R_i respectively, via an approximate $45°$ degree log transformation:

$$m_i = \log(R_i) - \log(G_i), \tag{4.5}$$

$$a_i = \frac{1}{2} \left(\log(R_i) + \log(G_i) \right). \tag{4.6}$$

Figure 4.10(b) shows an example of such transformation. The next step involves finding a smooth curve through the points. Historically, scatter plot smoothers, such as loess, have been suggested, but others such as smoothing splines work as well. From a practical point of view, it is important that they are implemented in one's package of choice. In R, the functions loess and smooth.spline are implemented. Like all smoothing functions, they require a smoothing parameter, that is, the amount of smoothing that needs to be applied. For the loess function, this parameter has to be supplied. Yang et al. (2002a) report that a span $= 0.2$ is a sensible choice. We add that the degree of the loess function is best set to 1, in order to avoid excessive smoothing effects at the edges. The smooth.spline function has the ability of selecting a sensible parameter by itself. We have had mixed results using it.

The smoother function can be applied to the whole data set or a subset thereof. In particular, if there is a subset of probes that is known to be invariant under the treatments in the Cy3 and Cy5 channel, then the dye normalization can use only those genes. However, it is important that the set of invariant genes is rather large in order to counterbalance any natural variation in these probes that are not due to the dye.

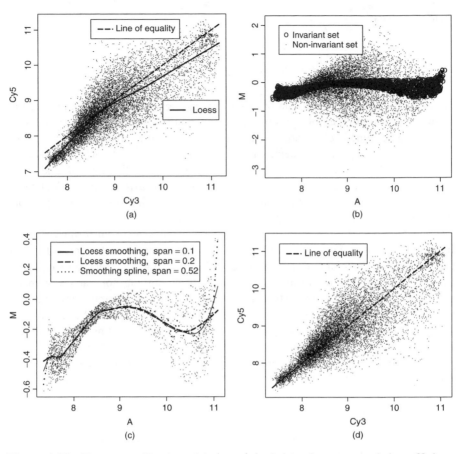

Figure 4.10 Dye normalization: (a) the original data show unequal dye efficiencies; (b) the data is transformed on the MA scale and a set of invariant genes between the two channels is determined (preferably on the basis of some biological criterion); (c) for those invariant genes a smoothed line is fitted to the scatter plot; (d) the residuals of the smoothed regression are back-transformed to the original scale.

Subsequently, the residuals of the data with respect to the smoothed line $\tilde{f}$ constitute a normalized MA plot,

$$\tilde{m}_i = m_i - \tilde{f}(a_i), \tag{4.7}$$

$$\tilde{a}_i = a_i. \tag{4.8}$$

By taking the inverse of Equations (4.5) and (4.6) using the normalized MA values in Equations (4.7) and (4.8), we obtain the normalized red and green channel values:

$$\log \tilde{R}_i = \tilde{a}_i + \tilde{m}_i/2, \tag{4.9}$$

$$\log \tilde{G}_i = \tilde{a}_i - \tilde{m}_i/2. \tag{4.10}$$

Numerical example.   The data from the first skin cancer array in Figure 4.10(a) show a clear non-constant dye effect. No information about spiking controls or housekeeping genes, which are expected to be similarly expressed across the two dyes, is available. One can decide to use the whole data set for normalization, but instead, the data are trimmed to eliminate those genes that have dramatically altered expressions across the two dyes. An *ad hoc invariance set* is defined as those genes whose relative rank among all 9,216 values has not changed by more than 300. In this fashion 2,365 probes are selected.

The Cy3 and Cy5 values for the invariant probes are transformed according to Equations (4.5) and (4.6). Figure 4.10(b) shows the transformed values for this array. The invariant set is used to calculate the smoothed curve of equality. Figure 4.10(c) shows the results of two different smoothing approaches, namely, `smooth.spline` and `loess`. The latter is shown with two different parameter settings. The smoothing spline is too variable, especially at the edges. We recommend using a loess smoother with a span of at least 0.2 and consisting of local linear lines (`degree = 1`). After subtracting the loess line from the MA-plot, the data are transformed back to the original scale, as shown in Figure 4.10(d). The data now do not show any intensity-dependent deviation from the line of equality.

The `dye.norm` function in R.

The loess-based, intensity-dependent dye normalization procedure has been implemented in R as the function `dye.norm`. It takes the `cy3` and `cy5` channel inputs and applies on all or a subset of the data the intensity-dependent loess smoothing with default `span = 0.2`. If this `invariantset` is specified as NA, it will take all the data; if it is specified as a single number, it will take all the spots whose expression rank has not changed by more than that number; or if `invariantset` contains a vector of numbers, it will take the subset of invariant spots as the data rows corresponding to the numbers in the vector `invariantset`.

4.3.5   Normalization within and across conditions

All the normalizations discussed so far have only dealt with the normalization of a single array. By dealing with spatial effects and dye effects, we hoped to achieve inter-comparability of the data from anywhere on the array in each of the channels. However, there are good reasons to suppose that data from replications *across* different arrays cannot be directly compared. For example, the preparation of the hybridization sample depends on the handling of liquids and dyes in small quantities where relatively large levels of variation can occur. In addition, scanning can occur at different gains and different photo multiplication levels. All this makes it necessary to position the data from different arrays on the same scale first, before attempting to analyse them.

Global method to normalize across arrays

The best-known method to deal with the different measurement scales from different arrays has been proposed several times under different names. Effectively, these methods transform the data by bringing the mean or median intensity for each array to some fixed quantity and possibly by adjusting the scale to another fixed value. Kerr et al. (2000) and Wolfinger et al. (2001) perform the location normalization by means of including an *array effect* into a factorial model. The location and scale normalization in Yang et al. (2002b) achieves relatively similar scales across replicates, although it simultaneously also attempts to account for any spatial effects on the array. In the single-channel literature, Affymetrix (Affymetrix 2001) introduced scaling factors to set the mean expression across all the arrays in an experiment to 100.

This normalization can be described as follows. Assume that one has data for p spots for the two dyes over n arrays, represented by

$$x_{ijk}, \quad \text{where } i = 1, \ldots, p, \quad j = 1, 2, \quad k = 1, \ldots, n.$$

Let m_{jk} and s_{jk} be some location and scale parameter respectively, for the jth dye across the kth array. Then,

$$x_{ijk}^* = \frac{x_{ijk} - m_{jk}}{s_{jk}}, \quad i = 1, \ldots, p$$

are the normalized data for array k and dye j. Unfortunately, these data are not on the original scale, and therefore, it is sensible to transform them back to a more interpretable scale. By taking m and s to be some overall summary of m_{jk} and s_{jk} ($j = 1, 2, \ k = 1, \ldots, n$) respectively, the spot intensities are transformed back to the original scale via,

$$x_{ijk}^{**} = m + s \times x_{ijk}^*.$$

The resulting expressions across all the arrays will then be on the same scale, that is, will have the same location and scale parameter, which is close to the original scale.

The choice of location and scale parameter is largely arbitrary. The most common ones are the mean and standard deviation. We recommend, as a matter of course, using relatively robust quantities, such as the median and the median absolute deviation, respectively.

The main objection against this type of method is the assumption of linearity. This normalization supposes that the variations in brightness across the arrays can be described by a shift and perhaps some form of multiplicative shrinkage.

However, the observed intensities are certain to lie between some fixed values. For a 16-bit Tiff image, these values are 0 and $2^{16} - 1$. Arrays that are brighter tend to have compressed values near the top intensity range, whereas arrays that are darker possess a compression of values near the lower intensity range. How serious this non-linear compression is depends on the overall stability of the experiments.

Another objection against the global location and scale stabilization method is that it ignores possible array-wide changes across the array due to different conditions. For example, it is known that during lactation, the expression of many genes in the mammary gland are suppressed. If a chip with mRNA from a mammary gland during pregnancy is compared with a chip with mRNA during lactation, then it is expected that the second chip would show lower overall expression. However, the global normalization method will set the level of each of the two arrays at the same overall level. Whereas, this may not be a large problem for complete genome arrays, for small custom arrays with only a few hundred genes on it this is a relevant issue.

In the next two sections, we discuss a quantile normalization method that deals with both the problems.

Recommended within-replication normalization

A microarray experiment is often replicated several times in order to deal with the natural variation across the array. Given that the same genes are measured under the same condition, it is natural to expect that the results should somehow be similar. That is the intuition behind the location and scale normalization discussed in the previous section. It is clear, however, that frequently they are not identically distributed and that simple location and scale adjustments are not sufficient. Figure 4.11 shows two replicates of the skin cancer experiment after the spatial, background and dye normalizations described in the previous sections. Although they are replicates—in the sense that the hybridizations used mRNA coming from the same cancerous cell line—the data do not seem to display the same distribution, especially at the top end of the scale.

A quantile normalization method has been proposed by several authors to deal with this *caveat* (Bolstad et al. 2003; Wernisch et al. 2003). It involves transforming all the replicates onto the same scale. This scale is itself obtained by an overall mean of the distributions of all the replicates. Figure 4.12 shows how the method works schematically.

The method is able to deal with non-linear compressions because effectively it only takes the ranks of the observations into account. Moreover, by using all the replicates of a particular observation, it is hoped that the 'compromise distribution' is a better reflection of the underlying gene transcription. In particular, the 'compromise' scale might be more 'linear' than the original scale.

Mathematical details.   Quantile normalization is applied to a matrix $\{x\}$ of spot intensities for all p genes and k replicates. Let x_{ij} be the spot intensity for spot j

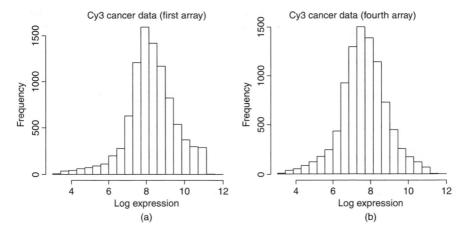

Figure 4.11 Even though they are replicates, the distributions of the data in the two arrays shown in (a) and (b) are not quite the same. This suggests that a simple location and scale normalization might not be sufficient.

on array i. Also, define the following two quantities,

$$x_{(j)} = \text{vector of } j\text{th smallest spot intensities across arrays,}$$

$$\overline{x}_{(j)} = \text{mean/median of } x_{(j)}.$$

The vector $(\overline{x}_{(j)})_{j=1,\dots,p}$ represents the 'compromise' distribution. Let $\{r\}$ be the matrix of row ranks associated with matrix $\{x\}$. Then,

$$x_{ij}^{\text{norm}} = \overline{x}_{(r_{ij})}$$

are the quantile normalized values of $\{x\}$.

As a numerical example, we associate some values to Figure 4.12. Let the following matrix represent two array replicates of five spots,

$$x = \begin{array}{c|ccccc} & 1 & 2 & 3 & 4 & 5 \\ \hline 1 & 16 & 0 & 9 & 11 & 7 \\ 2 & 13 & 3 & 5 & 14 & 8 \end{array}$$

The associated matrix of ranks is given as,

$$r = \begin{array}{c|ccccc} & 1 & 2 & 3 & 4 & 5 \\ \hline 1 & 5 & 1 & 3 & 4 & 2 \\ 2 & 4 & 1 & 2 & 5 & 3 \end{array}$$

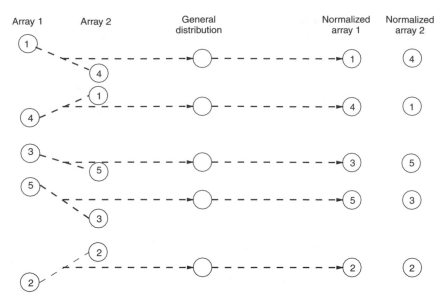

Figure 4.12   Schematic representation of quantile normalization within a certain condition.

The vector of the compromise distribution is obtained by averaging out the quantiles of the rows of x, that is,

$$\bar{x} = (1.5, 6, 8.5, 12, 15).$$

Then, mapping the ranks r onto this vector $\bar{x}$ yields the quantile normalized values,

$$x^{norm} = \begin{array}{c|ccccc} & 1 & 2 & 3 & 4 & 5 \\ \hline 1 & 15 & 1.5 & 8.5 & 12 & 6.0 \\ 2 & 12 & 1.5 & 6.0 & 15 & 8.5 \end{array}$$

This normalization can be considered a success if the overall variation of the spot replicates has not increased. In this example, we see that this is the case for all spots except spot 5.

Recommended across-condition normalization

Although it is possible to extend the quantile normalization across different conditions, there may be a biological reason why this is inadmissible. Although it is reasonable to assume that the distributions of two replicates under the same condition are approximately the same, it could well be that genome-wide changes may have occurred as a result of a different condition. This would affect the distribution of spot values across the array. In particular, this would be the case when the array contains only a few genes that have been selected on the basis of this expected response to the conditions.

As a result, one has to be more careful when normalizing replicates from different conditions. Only genes that are known to not change between conditions should be used in the normalization across conditions. In particular, just as for the normalization of the dye effect, a set of invariant genes is needed.

We propose a variation of the quantile normalization method across several conditions that can be graphically represented as in Figure 4.13. This form of quantile normalization performs an ordinary quantile normalization across the invariant genes, and then for each array fills in the remaining genes by linear interpolation. Interpolation only works if the smallest and the largest values on each array are part of the 'invariance set.' Therefore, by default they are added to the set of invariant genes.

The method is internally consistent, in that if the whole set of genes is declared invariant, it reduces to the ordinary quantile normalization method.

Mathematical details.   Let the data, again, be represented as a matrix $\{x\}$ of spot intensities for all p genes and k arrays. Let x_{ij} be the spot intensity for spot j on array i. The rows of the matrix now correspond to several different conditions.

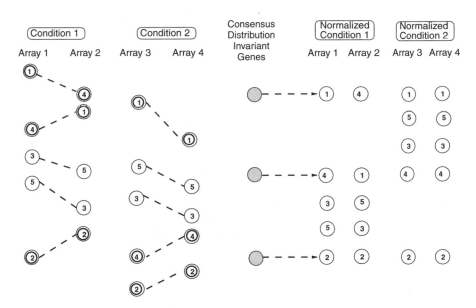

Figure 4.13   Schematic representation of a quantile normalization across several conditions, where genes 1 and 2 are not expected to change across the conditions. In order to avoid extrapolation, gene 4, which is the maximum on the second array, is added to the set of 'invariant' genes.

The point of departure for the quantile normalization method across conditions is the set of invariant genes. This set should be chosen on the basis of biological considerations and not on the basis of the data. Let

$$\mathcal{N} = \text{genes } a \text{ priori unaffected by the conditions.}$$

Define also the set of minima and maxima,

$$\mathcal{M} = \text{genes with smallest or largest expression in an array.}$$

For the moment, we restrict our attention to the data matrix $\{y\}$ of all the invariant genes and the genes that appear as minima or maxima,

$$y = [x_{n_1}, \dots, x_{n_{|\mathcal{N}|}}, x_{m_1}, \dots, x_{m_{|\mathcal{M}|}} \mid n_i \in \mathcal{N}, m_j \in \mathcal{M}].$$

Define the following two quantities,

$$y_{(j)} = \text{vector of } j\text{th smallest spot intensities across } y,$$

$$\overline{y}_{(j)} = \text{mean/median of } y_{(j)}.$$

The vector $(\overline{y}_{(j)})_{j=1,\dots,p}$ represents the 'compromise' distribution for the invariant genes. Let $\{r^y\}$ be the matrix of row ranks associated with matrix $\{y\}$. Then,

$$y_{ij}^{\text{norm}} = \overline{y}_{(r_{ij}^y)}$$

are the quantile normalized values of the invariant genes. In order to obtain the normalized values for all genes across all arrays, linear interpolation is performed. Let j' and j'' be two genes in $\mathcal{N} \cup \mathcal{M}$ that are the *nearest* invariant genes to x_{ij}, such that

$$x_{ij'} < x_{ij} < x_{ij''}.$$

Let y_1 and y_2 be the quantile normalized values associated with $x_{ij'}$ and $x_{ij''}$ respectively. Let there be a values between $x_{ij'}$ and x_{ij}, and b values between x_{ij} and $x_{ij''}$ on array i. Then

$$x_{ij}^{\text{norm}} = y_1 + \frac{a+1}{a+b+1}(y_2 - y_1)$$

is the quantile normalized value of gene j on array i.

As a numerical example, we associate some values to Figure 4.13. Let the following matrix represent two array replicates for each of two conditions over five spots,

		1	2	3	4	5
	1	16	0	9	11	7
$x =$	2	13	3	5	14	8
	3	13.5	−2	6	1	8.5
	4	10	−1	4	3	7

The set of invariant genes is $\mathcal{N} = \{1, 2\}$, while the set of minima and maxima across the arrays contains $\mathcal{M} = \{1, 2, 4\}$. For the set $\mathcal{N} \cup \mathcal{M}$ the associated matrix of ranks is given as

$$r^y = \begin{array}{c} \underline{1\ 2\ 4} \\ 1\ 3\ 1\ 2 \\ 2\ 2\ 1\ 3 \\ 3\ 3\ 1\ 2 \\ 4\ 3\ 1\ 2 \end{array}$$

The vector of the compromise distribution is obtained by taking the medians over the quantiles of the rows of x, that is,

$$\bar{y} = (-0.5, 7, 13.75).$$

Then, mapping the ranks r^y onto this vector $\bar{y}$ yields the quantile normalized values for the set $\mathcal{N} \cup \mathcal{M}$,

$$y^{\text{norm}} = \begin{array}{cccc} & 1 & 2 & 3 \\ 1 & 13.75 & -0.50 & 7.00 \\ 2 & 7.00 & -0.50 & 13.75 \\ 3 & 13.75 & -0.50 & 7.00 \\ 4 & 13.75 & -0.50 & 7.00 \end{array}$$

The remaining two genes are inserted for each array using interpolation:

$$x^{\text{norm}} = \begin{array}{ccccc} & 1 & 2 & 3 & 4 & 5 \\ 1 & 13.75 & -0.50 & 4.50 & 7.00 & 2.00 \\ 2 & 7.00 & -0.50 & 2.00 & 13.75 & 4.50 \\ 3 & 13.75 & -0.50 & 9.25 & 7.00 & 11.50 \\ 4 & 13.75 & -0.50 & 9.25 & 7.00 & 11.50 \end{array}$$

Figure 4.14 shows how the quantile normalization succeeds in aligning the distributions.

By normalizing first *within* condition before normalizing across conditions, it might be feared that the quantile normalization method introduces spurious differences between conditions. We performed a simulation study with ten genes and two conditions to test this hypothesis. Each of the conditions was replicated four times. None of the genes was assumed to be differentially expressed and all the gene expressions were taken from independent normal distributions with mean $= 0$ and

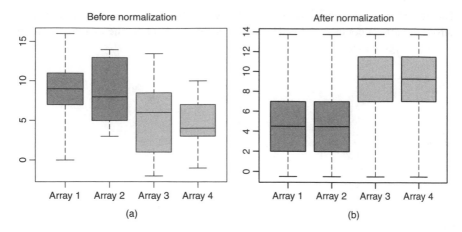

Figure 4.14   (a) Simulated data, consisting of five expression values over four arrays across two conditions; (b) data after the quantile normalization across the invariant genes.

Table 4.4   Observed significance level ($\frac{\text{\# rejections}}{10 \times 500}$) over 500 replications of 10 non-differentially expressed genes, using a t-test at $\alpha = 0.05$ over (i) the raw data, (ii) quantile normalized data and (iii) quantile normalized data with $\mathcal{N} = \{1, 2\}$.

	(i)	(ii)	(iii)
$\hat{\alpha}$	0.041	0.042	0.041

standard deviation = 1. Besides using the raw data to test for differential expression, we applied the two different quantile normalizations to the data: (i) quantile normalization across all genes, (ii) quantile normalization with invariance set $\mathcal{N} = \{1, 2\}$. In all the cases, we used the ordinary t-test at a $\alpha = 0.05$ significance levels to test for mean differences between the expressions. Table 4.4 suggests that no systematic bias is introduced by the methods we propose. However, we have noticed that methods that actively force distributions within conditions to be the same might introduce bias (simulations not shown).

The `cond.norm` function in R.

The quantile normalization method within and between conditions has been implemented in R as the function `cond.norm`. It relies on being presented as a data matrix, where the rows represent the arrays and the columns represent the spots.

If it does not receive any further information, then it proceeds to do a quantile normalization on all the genes and arrays simultaneously. The result is that all the arrays end up on precisely the same scale. If information about invariant genes is entered in `invariantset` as a set of column numbers or as a logical vector, then `cond.norm` only employs those spots to quantile normalize the data. By supplying information about the `conditions`, the data are first quantile normalized across replicates, before the across-condition normalization. If the `method` is changed from `quantile` to `global`, then a simple location and scale normalization is applied to all the arrays.

4.4   Normalization of Single-channel Arrays

Single-channel arrays, in principle, do not produce any normalization obstacles that have not been already encountered during the normalization of dual-channel arrays. For single-channel arrays there can be spatial effects, background noise and general scale differences between arrays. All the normalization techniques discussed in Section 4.3, with the exception of the correction for the dye effect, apply to single-channel arrays. The most common single-channel microarray, the Affymetrix GeneChip, has some idiosyncrasies when it comes to data storage and initial analyses, which justifies a normalization section of its own.

An Affymetrix GeneChip comes with internal replicates for each gene. Typically, 11 to 20 so-called probes are present on each array. Each of the probes are spots with small, gene-specific oligonucleotides, 25 base pairs long. An oligonucleotide can be considered a brief quote of a particular gene. The idea is that as long as the quote is specific enough, RNA from that and only that gene can recognize it as itself.

However, undoubtedly, RNA from other genes may find parts in common with the individual probe and attach itself to it. To deal with this problem of non-specific hybridization, Affymetrix devised for each probe a so-called 'mismatch' cousin that was obtained by changing the 13th base pair of each oligonucleotide probe. It was believed that by subtracting the MM value from the observed 'PM' probe values, a gene-specific expression value would be obtained.

4.4.1   Affymetrix data structure

Data and information about the data from Affymetrix GeneChips are stored in a variety of ways. Table 4.5 gives a complete overview. For our purposes, the most important files are the CEL (`*.cel`) and CDF (`*.cdf`) files. The CEL file contains the information about the individual probes. This file is the result of the scanning and segmentation of the subsequent image. However, it has not yet summarized the probe values into individual gene expression values. The CEL files contain the average probe pixel intensities, the probe pixel standard deviations and also information about the physical location of each probe on the array. For each array there exists a corresponding `*.cel` file. To make biological sense of the CEL files,

Table 4.5   Affymetrix data information files.

Data files	
*.exp	experimental information file
*.dat	image file
*.cel	probe intensity file
*.chp	gene intensity file
*.rpt	report file
Probe information files	
*.cif	chip information file
*.cdf	chip description file
*.msk	mask file

a map from the probes to the gene names is needed. This information is stored in the CDF file. Unless different GeneChip types were used, only one CDF file is needed for an experiment that consists of several arrays.

Reading in Affymetrix data into R

The affy package is an add-on package for R written by Irizarry et al. and described in Irizarry et al. (2003a). This package allows the user to read Affymetrix files into R without having to convert the files first. The commands,

```
> objectname.cel  <- read.celfile("filename.cel")
```

and

```
> objectname.cdf  <- read.cdffile("filename.cdf")
```

read in the CEL and CDF files respectively.

4.4.2   Normalization of Affymetrix data

Several methods have been proposed to normalize Affymetrix data. Affymetrix's own MAS 5.0 software subtracts the MM values from the PM values to get the individual probe expression values. Some *ad hoc* adjustments prevent the difference from becoming negative. By using a relatively robust location estimate called *Tukey's biweight*, the probe values for a particular gene are summarized into a single gene expression value. A scaling factor makes sure that all arrays have a similar average brightness.

   Li & Wong's dChip project (Li and Wong 2003; Li et al. 2003) normalizes the data by pooling for each gene the information across probes and across replicates using a multiplicative model. Their method automatically detects and removes probe or array outliers. However, it does not normalize for spatial or other effects, except for standardizing the overall brightness across the arrays.

Irizarry et al. (Irizarry et al. 2003a,b) suggest a novel background correction in the place of Affymetrix's mismatch subtraction. The associated R library, `affy`, includes their background correction method as well as Li & Wong's dChip method. However, again, no adjustments are made for varying levels of brightness across the array.

In this section, we describe the methods by which we propose to deal with the most common sources of unspecific variation in Affymetrix data—spatial variation, background noise and overall differences in brightness between arrays. All of these methods have been described before in the context of the dual-channel arrays. The R functions that were proposed for the dual-channel arrays also work for the Affymetrix data. An effort has been made to minimize duplication and interact with the `affy` package.

Spatial correction

Just like a dual-channel array, Affymetrix GeneChip is an optical technique whereby gene expressions are transformed into optical intensities, which can be read by a scanner. Images are notorious for containing spatial effects. There might be correlations induced in the scanning process itself, or, more likely, the spatial dependencies in the hybridization sample can lead to uneven brightness across the array. Figure 4.15 shows the subtle variations of the intensities across an Affymetrix array on the log scale. Fluctuations between 5.35 and 5.65 on the log scale stand for a 15% fluctuations either way on the original scale. This is a physically significant value and typical for the type of fluctuations that can be observed. Figure 4.15 also shows array damage at different places. The best thing would be to take these values out and consider them as missing values. Imputation of missing values is discussed in Section 4.2.3.

Use of the `spat.norm` function for Affymetrix data in R. The same function that is used for estimating the brightness fluctuations across the dual-channel array (*cf.* Section 4.3.2) can be used to normalize the spatial effect on Affymetrix arrays. To extract the probe intensity data from the cell object that we obtained by using the `read.celfile` function in the `affy` package in R, we have used the function `cel2int`. It returns a matrix with probe intensities on the original GeneChip grid. This matrix can then be directly entered into the `spat.norm` function. To make the normalization robust to outliers and left skewed data in general, we recommend normalizing the log-transformed data,

> mydata.sn < −spat.norm(log(cel2int(mycel.object))).

The function will return a matrix of the original size with spatially normalized intensities. It is possible to put the spatially corrected data back into the cell object, via

> attributes(mycel.object)$intensity < −mydata.sn.

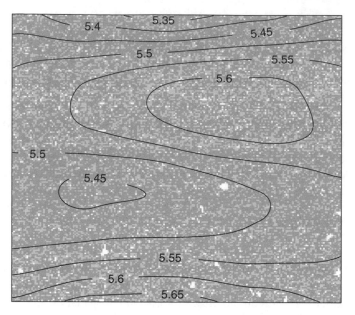

Figure 4.15   An Affymetrix array with superimposed contour lines of the smoothed surface.

Affymetrix arrays tend to have many more probes on each array than common
printed arrays. This makes the spatial normalization slower. However, by normalizing only across a large random sample of points, indicated by n.sample, the
normalization speeds up significantly. The current default is n.sample = 10,000,
which corresponds to the number of probes of an ordinary cDNA microarray.

Background correction

Image analysts are used to dealing with background noise in images. This is a
random but possibly spatially correlated noise that adds itself to the location-
specific signal. Affymetrix has developed its own ways to deal with non-specific
or so-called *stray* signals (Hubbell et al. 2002). In particular, in MAS 5.0, as
suggested by Hubbell et al., the true signal T_j for probe j is estimated as

$$T_j = \mathrm{PM}_j - p\mathrm{MM}_j,$$

where $p = 1$ if the PM signal (PM_j) is larger than the MM signal (MM_j), or p is
an *ad hoc* value between 0 and 1 if the PM signal is less than the MM signal, so
that T_j is always positive.

Besides the fact that the choice of p is *ad hoc* and not informed by any physical
model, this method also suffers from very similar problems as those with local

background subtraction, discussed in Section 4.3.3. In particular, as the mismatch, like any other quantity, is observed with error, subtracting it from the PM value tends to add additional noise. Moreover, gene-specific RNA also hybridizes to the mismatch, and therefore, subtracting it is precisely the incorrect thing to do (Irizarry et al. 2003b).

Instead of subtracting the MM values from the PM values, several other options are open to deal with unspecific hybridization. One can consider looking at the intensity of empty spots and subtract an average value of these spots from each probe value. The background normalized value x_j^* for probe j is then defined as

$$x_j^* = \max(x_j - \mu_{\text{empty}}, 0). \tag{4.11}$$

Another method is to consider the distribution from the empty spots as an indicator for the amount of noise in its probe. By making a few distributional assumptions on the stray signal and true signal, a conditional expectation of the true signal can be calculated:

$$x_j^* = E(t_j | x_j = t_j + b_j), \tag{4.12}$$

where t_j and b_j are the true and background signal for probe j respectively. This is the so-called *probabilistic background correction method*, also discussed in Section 4.3.3 and in Irizarry et al. (2003b). It is our recommendation to use the background correction sparingly. In Figure 4.16 background correction was applied to a pair of sets of Affymetrix data. The data came from two scannings of the same array. It shows that on the log scale the background corrected intensities are more variable than the original data. This is a bit deceptive, since the background correction really takes place on the original scale. Nevertheless, it is our experience that, of all the normalization methods, background correction is the one that least reduces the variation between replicate arrays. This is a sign that it does not correct for much of this particular source of variation.

Use of the bkg.norm function for Affymetrix data in R. The background correction function bkg.norm takes as its main argument a vector of probe intensities. The option empty can be used to specify the locations of empty spots. If a number n is provided, it simply assumes that the lowest n probe values come from empty spots. It is also possible to enter the mean and standard deviation information of the empty spots manually. These values will then be used in the background correction.

The function can perform both the deterministic correction in Equation (4.11) or the probabilistic correction in Equation (4.12) by setting the method option to "deterministic" or "probabilistic" respectively. The probabilistic background normalization function is also implemented in the affy library. The argument bg = rma in the function express performs this normalization.

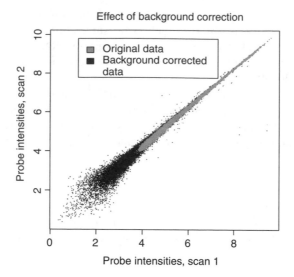

Figure 4.16   Original data seems to show less variation than the background corrected data.

Correction within and across conditions

The overall brightness of an Affymetrix array can vary from chip to chip. Arrays might be scanned at different photomultiplier levels and different gains. The hybridization mix may have been of different concentrations. It is therefore important to bring different arrays onto the same scale. Affymetrix MAS 5.0 package (Affymetrix 2001) proposes to use scaling factors to ensure that each array has the same mean or median probe intensity. This corresponds to the global scaling method discussed in Section 4.3.5. The main disadvantage of this method is that it is unable to deal with non-linear distortions at the lower and higher end of the probe intensity scale. These distortions are the result of the fixed interval $[0, 2^{16} - 1]$ in which all of the probe intensities are recorded.

   Another method is the so-called quantile normalization. It forces the distribution of all the replicate arrays to follow some kind of 'average' distribution. An example of this can be seen in Figure 4.17. After the quantile normalization, the two replicates line up precisely along the line of equality.

   Although this type of normalization is useful when two arrays are replicates, it might introduce bias when the arrays represent different biological conditions. In such case, it is not necessarily the case that most of the probes are expected along the line of equality. For this purpose, it is important to define a set of probes that are expected to remain invariant under the conditions. By applying the quantile normalization only to the invariant set and interpolating all the other probes, biological information is taken on board to avoid introducing bias by over-normalizing the data. The precise details of the algorithm are discussed in Section 4.3.5.

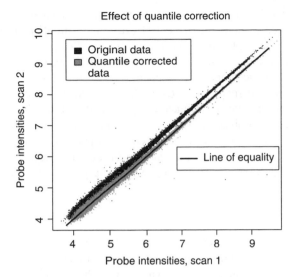

Figure 4.17   Quantile normalization brings the two technical replicates on the same scale.

Use of the `cond.norm` function for Affymetrix data in R.   The quantile normalization is implemented in the R function `cond.norm`. It takes as argument a matrix of probe values, whereby the rows are the arrays and the columns are the genes. By specifying the samples via the `conditions` argument and the invariant genes via the `invariantset` argument, a complete normalization takes place within conditions and a partial normalization across conditions. Further details of this algorithm can be found in Section 4.3.5.

The function `cond.norm` can also be instructed to perform a global normalization, whereby the median and median absolute deviation (MAD) are made equal across all arrays.

Summarizing probe set data into gene expression values

Each gene on the Affymetrix GeneChip is represented by a number of probes, typically between 11 and 20. This group of probes is called a *probe set*. The probes themselves are different 25 base pair 'quotes' from the gene of interest. Whereas before, the probes of a gene were put next to each other, it was soon recognized that this introduced needless opportunities for spatial bias. On modern Affymetrix chips, the probes of a single gene are spread across the array.

Although it is useful to have many probe values, they should not be confused with biological replicates. The values are associated with a single hybridization of a single sample. Therefore, it is statistically prudent to summarize the probe values into a single gene expression value, as there is only one degree of freedom.

The simplest method is to take a (trimmed) average. Affymetrix implemented this in its MAS 4.0 software as the *AvgDiff* value. It was observed that some probes were much brighter than others. In response, Affymetrix changed this in MAS 5.0 and used a more robust central estimate. Although the median would probably suffice, they opted for a more complicated *Tukey Biweight* estimate.

The observation that certain probes are fundamentally brighter than others belies the assumption that the probes are independent and identically distributed values from some common, perhaps heavy tailed distribution. If this assumption is really in doubt, it might be better to estimate the gene expression value by allowing the effect of the individual probes to vary. This lies behind the so-called 'model-based analysis of oligonucleotide arrays' proposed in Li and Wong (2000) and Li et al. (2003). The *analysis* is a way to estimate the probe efficiency ϕ_j for probe j and the *model-based expression index* (MBEI) θ_i for a particular gene on array i via the model:

$$y_{ij} = \theta_i \phi_j + \epsilon_{ij}, \quad i = 1, \ldots, n_{\text{arrays}}, \quad j = 1, \ldots, n_{\text{probes}}.$$

where y_{ij} is the observed probe intensity on the original scale. This is a multiplicative model on the non-logged scale. One of the advantages of this method is the ability to detect unusual probe values y_{ij}—having high leverage or large standardized residuals—which can be eliminated from the data automatically by setting some tuning parameters.

Table 4.6 shows the approximate relationship between methods that take (complicated) averages of the probe values and those that allow for an individual probe effect. Irizarry et al. (2003a) propose a variation on the theme. By transforming the multiplicative model on the original scale into an additive model on the log scale, the two-way model is approximately equivalent to the Li & Wong method and much easier to estimate. Also, they allow for an individual probe effect, but rather than fitting a multiplicative model, they propose a related additive model on the logarithmic scale,

$$\log(y_{ij}) = \theta_i + \phi_j + \epsilon_{ij}, \quad i = 1, \ldots, n_{\text{arrays}}, \quad j = 1, \ldots, n_{\text{probes}}.$$

Traditionally, this model was fitted by least squares, but today there are many other methods to fit this model more robustly, such as median polish or robust regression

Table 4.6 Main approaches for estimating the gene expression value θ_i from several probe values j over different samples i: Model 2 (Irizarry's RMA and, approximately, Li & Wong's model-based expression index) is an extension of the one-way model, such as taking simple averages (*AvgDiff* method).

	Method	Formula
1.	One-way ANOVA	$\log(y_{ij}) = \theta_i + \epsilon_{ij}$
2.	Two-way ANOVA	$\log(y_{ij}) = \theta_i + \phi_j + \epsilon_{ij}$

using an M estimator (Venables and Ripley 1999). Irizarry et al. (2003a) use median polish to derive estimators for the expression measures θ_i.

Implementation in R.   The different probe summary methods are implemented in the `affy` package in R. The argument `summary.stat` of the function `express` controls which summary method is selected and can be set to `avdiff` for taking a (trimmed) mean, to `li.wong` for the Li-Wong expression index, or to `rma` (default) for the additive method by Irizarry et al.

5

Quality Assessment

Although the advances in technology over recent years have been vast and microarray analysis has become a commonly used tool, microarray experiments are complicated and many sources of unwanted variation are present when conducting them. Array normalization can and should be used to remove many of these noise components in the data. However, many sources of variation may be present in a microarray experiment, and normalization cannot be expected to correct all of them. Further, it is always worth checking whether the normalization has satisfactorily accounted for the problems with the arrays. It is also worth checking the qualitative information about an experiment, such as that contained in MIAME.

This chapter deals with methods for determining when and which arrays may be unreliable. It does not and cannot be exhaustive as there can be many types of problems, but the methods given carry out important checks on the suitability of the data. The main types of problems that these checks can detect are the mislabelling of arrays, array-wide hybridization problems, problems with normalization and mishandling problems. We look at each of them in turn.

Four main techniques are used to identify these problems—dimension reduction techniques, which include visualization methods, pairwise scatter plots, false array images and correlograms.

Some work on quality assurance of microarray data is starting to appear (McClure and Wit 2003). Most imaging software programmes (e.g. Agilent Feature Extraction Software, MAS 5.0, QuantArray) have the ability to flag spots if the individual pixel values are too variable. These flags tend to be based on relatively crude and simple hypothesis tests. Li and Wong (2001) propose a probe-treatment model for Affymetrix gene expression. As a result, they are able to flag unreliable probes by means of an outlier detection method. This method has been implemented in the dChip package (Li and Wong 2003; Li et al. 2003).

Statistics for Microarrays: Design, Analysis and Inference E. Wit and J. McClure
© 2004 John Wiley & Sons, Ltd ISBN: 0-470-84993-2

5.1   Using MIAME in Quality Assessment

MIAME, or minimum information about a microarray experiment, is an attempt to standardize the storing of information about microarray experiments. It is the creation of microarray gene expression data (MGED) Society (www.mged.org), which includes many important microarray developers and users. The primary motivation behind MIAME is to create a standard for the storage of microarray information to enable the creation of useful, searchable databases.

Despite being labelled 'minimum', there is a great deal of information about an experiment within the MIAME, and some of this can potentially be used for quality assurance. What can be used is primarily dependent on what elements of the experiment change across the different arrays. For example, the quantity of mRNA may not have been as great in one sample compared to others, due to experimental error.

MIAME will tend to be used when assessing others' work, since it contains a lot of the information that would normally be available directly to experimenters or their collaborators. In this case, MIAME can be very useful for conducting *qualitative* quality assessment. When considering someone else's experiment the information contained on the protocols and equipment used can be very useful for deciding whether the experiment is worth investigating further.

5.1.1   Components of MIAME

Array design description.   This contains physical information on the arrays in the experiment. The nature of the arrays used can give insights into the performance of a microarray experiment. There are five main parts to the information recorded by MIAME:

1. Information is given on the array—its design name, platform type (synthesized, spotted, etc.), the surface and coating specifications, the size and number of spots, specified as *features*, on it. The availability or production protocol of an array can also be included.

2. Information is given on each distinct nucleic sequence, or *reporter*, on an array—the type, the sequence used and its length. Clone information, if relevant, can be recorded, along with the element generation protocol.

3. The type, size, attachment, location and respective reporter of each feature is also recorded.

4. Information about each *composite sequence* of features relating to a particular gene exon or splice variant is included—the reporters contained, the reference sequence and the gene's name.

5. For any control elements on the array, their position, type and exogenous or endogenous state is recorded.

Given that this is likely to be the same across all arrays used in an experiment, this information may not be of use for quality assessment.

Experiment description.   Besides its physical elements, an experiment is defined by how it is organized. The procedures used in the implementation, by whom and how they were followed—all these give important information on the quality of a microarray experiment. These procedures include the following:

1. Experimental design—information on the authors, type of experiment, experimental factors and replication.

2. Samples used—information on the samples includes the sex, age, development stage, genetic variation and disease state.

3. Extraction method—details of what manipulations are carried out on the samples, possibly including growth conditions, and treatments (*in vivo* or *in vitro*) are given. The method for the extraction of nucleic acids, their labelling and any spiking controls are also included.

4. Hybridization procedures—details of how the target is hybridized to the slide, including the protocols and which sample is applied to which solution.

5. Measurement data and data processing—raw images of each hybridization, image quantitations and normalized expression values, each with information on how they were obtained.

All this information can be very useful in assessing the quality of a microarray experiment. Further information is given about the elements in a MIAME description at the MGED website (MGED 2002).

5.2   Comparing Multivariate Data

A way to assess the quality of microarray data is by testing the internal consistency of the data. This section describes ways to compare data from a microarray experiment internally. It discusses the scale on which to measure gene expression data, how to assess the dissimilarity between gene expression profiles and how to visualize these dissimilarities.

5.2.1   Measurement scale

Microarray slides are generally stored in computer memory as 16-bit tiff files. Pixel values, therefore, range from 0 to $2^{16} - 1$, and as a consequence, spot mean or median intensities are almost always on a scale from 0 to 65,535. The intensities tend to be positively skewed so that most genes have an intensity much less than 65,535, with a few forming a 'tail' of increasing intensity extending towards 65,535.

Normalizations often change the scale of the data. Kerr and Churchill (2001c) and others suggest changing the data to the log scale because that turns multiplicative effects into additive ones and makes subsequent analysis easier. Rocke and Durbin (2001) and Huber et al. (2002) make a strong case that not only multiplicative effects are present but also additive ones. They suggest, instead, the application of a generalized log transformation, which has the advantage of stabilizing variances across the array.

However, whatever scale normalized data is on, it is always worth checking your data for artifacts on different measurement scales. Different problems can affect the data in different ways, and considering the data on one scale can allow certain features to be detected easily, whilst obscuring others. So, even though it may not be the ultimate scale on which you wish to analyse the data, transforming your data after they have been normalized can be an excellent diagnostic tool to spot normalization problems.

A number of transformations can be used for doing this, and the ones to be used depend on the scale of the normalized data (McClure and Wit 2003). If the normalized values are on the original positively skewed scale, then taking logs can highlight problems at the lower end of the intensity scale. The original scale is more useful for picking up problems at high intensities. In general, it is useful to consider a parameterized class of transformations, such as the class of all power transformations,

$$p_\alpha : x \mapsto x^\alpha,$$

where x is the normalized intensity; note that $\alpha = 0$ stands for the log transformation. Higher powers ($\alpha \geq 1$) are particularly useful for pronouncing top-end artifacts, whereas lower powers ($\alpha < 1$) can be used to spot low-end artifacts. However, artifacts that lie in the busy middle range of spot intensities cannot be enhanced by any such transformation. For those kinds of artifacts, it is useful to use the rank transformation that replaces each spot intensity by the relative rank it occupies on the array. This transformation creates equal spacing between all spot intensities and is therefore extremely valuable to enhance contrasts between intensities that are extremely close.

5.2.2 Dissimilarity and distance measures

The type of comparison we wish to make between different microarrays determines the *distance* or *dissimilarity measure* that should be used. A distance measure puts a value on the dissimilarity between two arrays. A variety of different measures exist, and they measure different aspects of the dissimilarity between arrays. Choosing an inappropriate measure for the question at hand can lead to misleading conclusions.

Two types of similarity we consider are the absolute similarity of two sets of gene expression levels and the correlation of gene expressions between two arrays. The former is related to whether each gene has approximately the same level of expression on one array compared to another; mean distance–based measures can be used for this. The latter looks for coordinated changes from gene to gene across two arrays; correlation measures can be used here.

The correlation-based measure used here is

$$d_\rho(x, y) = 1 - \rho(x, y),$$

where ρ stands for the ordinary correlation coefficient. This means that a correlation close to one has a dissimilarity of zero, as you would expect. The correlation between X and Y is given by

$$\rho_{XY} = \frac{\text{cov}(X, Y)}{\text{SD}(X)\text{SD}(Y)},$$

where $\text{cov}(X, Y)$ is the covariance and $\text{SD}(X)$ and $\text{SD}(Y)$ are the standard deviations of X and Y respectively. The correlation is therefore the normalized covariance.

Mean distance–based measures give an idea of how close all the gene expression values are across different arrays. We consider the set of power distances

$$d_p(x, y) = \sqrt[p]{\sum_{i=1}^{n_{\text{genes}}} |x_i - y_i|^p}. \tag{5.1}$$

These vary in their robustness to outliers. The most robust of these distance measures is Manhattan distance, d_1. This is the sum of the absolute distances between equivalent spots on two different arrays. As the distances between equivalent spots are raised to higher powers, the measure becomes more sensitive to outliers. This might be a desirable feature when you wish to detect outliers. The measure d_2 is known as *Euclidean distance*. This is the most intuitive measure of distance between two objects in two or three dimensions. In a two-variable example, it is simply the length of the straight line that joins the two points representing the object's value on a two-dimensional graph with the two variables as the axes. An example is given in Figure 5.1, where the two objects have values 1 and 6 for variable 1, and values 1 and 4 for variable 2. The Euclidean distance between these two objects is given by the length of the (solid line) straight line joining the points (1,1) and (6,4)—6 units in this case.

In the example shown in Figure 5.1 Manhattan distance is given by the length of the dotted line ($5 + 3 = 8$ units). It is the distance between the two points if we can only travel in one direction (of variable 1 or variable 2) at any one time, as if we were navigating a grid system like the streets of Manhattan.

Distance measures and scale transformations can be combined. For instance, the correlation using the ranks of the data is known as the rank correlation. It is robust to deviations from linearity, which, in the light of non-linear dye effects, might be quite useful. In general, one chooses a scale to indicate the region of interest on which to focus, for example, the low-end or top-end of the measurement scale. Then one selects a distance measure based on either mean expression differences

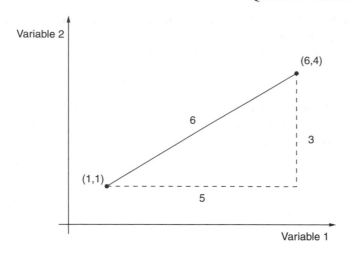

Figure 5.1   Example of Euclidean and Manhattan distance in a two-variable situation. The Euclidean distance between the objects at (1,1) and (6,4) is given by the solid line (6 units); the Manhattan distance is given by the dashed line (5 + 3 = 8 units).

or expression correlations. It is important to remember that the scale used for data comparisons affects the distances between arrays and that different scales can highlight different types of problems in the data.

Comparison of correlation and Euclidean measures.   The following example is used to illustrate the differences between mean distance based measures and correlation based ones. Figure 5.2 gives expression values of 20 genes on 3 different microarrays (labelled A, B and C), together with tables of correlation coefficients and Euclidean distances between these arrays. Note first that while larger correlations indicate a greater similarity, larger Euclidean and similar distance measures indicate greater dissimilarity. From the Figure and the tables, we can see that arrays A and B are very highly correlated (coefficient of 0.96), whereas A and C have almost no correlation. However, the Euclidean distance between A and B (136,266) is over 4 times greater than that between A and C (33,451), because array B has much lower expression values than arrays A and C.

The type of comparison between arrays that we wish to make will determine whether correlation-based or mean-based measures are better. If the mean expression of an array is important, then we should consider using a measure like Euclidean distance.

Comparison of Euclidean and Manhattan distance.   Figure 5.3 shows a plot with expression values for 20 genes on 3 arrays (labelled A, B and C) along with tables of the Euclidean and Manhattan distances between these arrays.

	Correlation				Euclidean ($\times 10^3$)		
	A	B	C		A	B	C
A	1	0.96	0.01	A	0	136	33
B	0.96	1	0.10	B	136	0	146
C	0.01	0.10	1	C	33	146	0

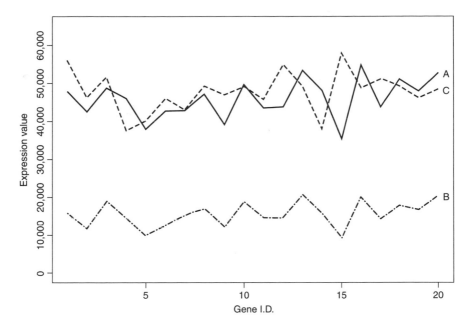

Figure 5.2   Expression values of 20 genes on 3 microarrays (labelled A, B and C), together with correlation coefficients and Euclidean distances between these arrays.

From Figure 5.3 we can see that, but for gene 14, arrays A and B appear much closer together than arrays A and C, as C appears to have a lower average expression. However, for gene 14, array B is extremely low—if all the arrays are replicates, something has probably gone wrong with the hybridization here. This outlier will obviously influence the distances between the arrays. The Euclidean measure is affected so much that arrays A and C appear more similar (with a dissimilarity of 40,187) than A and B do (where the dissimilarity is 43,441). Manhattan distance is not affected nearly as much by this outlier, and B is found to be more similar to A using this measure (dissimilarity of 93,307) than C is to A (dissimilarity of 174,015).

The Manhattan distance is more robust to outliers that often occur on otherwise well-hybridized arrays. This is likely to make it more suitable than Euclidean

Euclidean ($\times 10^3$)			Manhattan ($\times 10^3$)		
	A	B	C		A	B	C
A	0	43	40	A	0	93	174
B	43	0	47	B	93	0	186
C	40	47	0	C	174	186	0

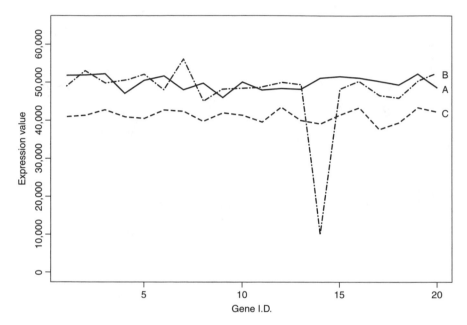

Figure 5.3   Expression values of 20 genes on 3 microarrays (labelled A, B and C), together with Euclidean and Manhattan distances between these arrays.

distance when comparing arrays. However, using both Euclidean and Manhattan distances would enable the detection of these outliers; if outliers were present, then the two distances would give different results.

What are the units of comparison?   This seemingly unnecessary question is nevertheless important. The type of array used and the type of check being conducted both determine what we treat as the entity of interest in quality assessment—what we consider to be a *unit of comparison*.

In the case of Affymetrix chips, this is simple—the array itself is the unit of comparison. With two-channel arrays, it is more complicated. As the two channels tend to represent different conditions, it will often be useful to treat the results from two channels for one array as separate comparison units. However, if multiple conditions are each tested against a reference on different arrays in an experiment, then some ratio of red to green may be the desired quantity. What is chosen as the comparison unit for a two-channel array will depend on the type of normalization done.

5.2.3 Representing multivariate data

Multivariate data can be compared by means of formal tests. However, the number of formal tests that are interesting are potentially infinite. Here, we focus on spotting potential problems in the data by eye. Visualization methods represent the multidimensional data, such as array-wide expression figures in a lower-dimensional space, generally a two-dimensional graph. These lower-dimensional summaries can be very useful for 'getting a feel' for the data.

Visualization methods are, however, *relative*. There are differences between any two arrays, no matter how similar they may be. Graphs produced by visualization methods adapt the range of their axes on the basis of the observed maximum difference, no matter how small that difference may be. In such cases, it may be hard to determine the importance of the differences.

We can do two things to overcome this problem. The first is to have as many replicates as possible. These replicates will be able to give an idea of what normal variation is supposed to be, and they allow one to gauge the level of unusualness of any possible outlier. The second way is to ensure that arrays of samples that have been subjected to different conditions are included. The benefit of having arrays that should be different from each other is, again, that it helps to create a yardstick for comparisons.

Dimension reduction methods

A number of different methods are possible for dimension reduction. We shall consider the very popular principal component analysis (PCA) and an alternative, known as Sammon mapping, that has certain advantages over PCA. The aim is to represent objects in a small dimensional space, so that their relative similarity can easily be seen. When conducting quality assessment of a microarray experiment, it is the arrays that are the objects and each gene is one of the original dimensions.

Principal component analysis. The most well-known dimension reduction method is principal component analysis (PCA). This can be a very useful way of reducing the dimension of multidimensional data. It has the advantage of being available in most standard packages. It can project the data into a two, three or greater dimensional space according to choice.

To understand PCA, we need to remember that each of the original dimensions is an axis. However, other axes can be created that are linear combinations of the original ones. This is what PCA does—it creates a completely new set of axes. Like the original axes, the principal components are all orthogonal to each other.

One of these components, known as the first principal component, is the axis through the data along which there is the greatest variation amongst the objects.

The second principal component is the axis orthogonal to the first that has the greatest variation in the data associated with it; the third principal component is the axis with the greatest variation along it that is orthogonal to both the first and second axes; and so forth. If there is correlation between the objects, then the first few principal components can account for a lot of the variation in the data; a plot of, say, the first two principal components can therefore be useful for looking for differences between objects.

However, there are some known problems with PCA. The first principal components may not minimize the within-cluster variance while maximizing the between-cluster variance. Also, the diminishing importance of subsequent principal components, which are used as axes, makes interpretation of a plot difficult. Moreover, implementation is based on the complete data matrix and can be rather slow.

Sammon mapping. Sammon mapping (Sammon 1969) uses the distance matrix between arrays rather than the original data matrix and hence is more quickly implemented than PCA or other dimension reduction methods. Sammon mapping aims to find a representation of the arrays in a lower-dimensional space in such a way as to minimize the total stress of the new representation with respect to the old one. It does so by finding new distances $\tilde{d}$ for a k-dimensional representation that minimize the weighted 'stress'

$$E(d, \tilde{d}) = \frac{1}{\sum_{i \neq j} d_{ij}} \sum_{i \neq j} \frac{(d_{ij} - \tilde{d}_{ij})^2}{d_{ij}}. \tag{5.2}$$

In other words, Sammon mapping finds the representation of the units of comparison that involves as little distortion of the distance matrix as possible. So, for the two-dimensional case, Sammon mapping calculates the two-way graph for which the distances between all the arrays are the closest to the equivalent distances in the original distance matrix. The 'stress' is the extent to which the distances are distorted by such a representation. Working on the distance matrix allows Sammon mapping to give a better representation of the arrays than PCA, especially if only a small fraction of the variation between the arrays is explained by the first two or three principal components.

False array plots

Having spotted a potential problem, a more detailed investigation of the array where this occurs is useful. One way of doing this is by creating a 'false array plot'. This is an image of the array using the normalized values to represent the intensity of each spot. If genes have been spotted on the array in a more or less random fashion, then these images should exhibit no patterns of any kind. Patterns that do stand out on these false array plots can give a clue of what could be wrong with a particular slide. Obviously, if there is a pattern to the spotting of the array, then this can confound any other pattern. This is one of the reasons why it is helpful if spots are allocated randomly to the array. These kind of statistical design issues are discussed further in Chapter 3.

Pairwise scatter plots of replicates

Although dimension reduction makes information more manageable, it does represent a loss of information. Not all types of quality problems can be spotted via dimension reduction. Multiple scatter plots may represent a useful compromise. Many authors have used scatter plots to evaluate the quality of their microarray data. For instance, Amaratunga and Cabrera (2001) use scatter plots of replicates to notice strange truncation effects at the top end of the intensity range. Dudoit et al. (2002) introduced a so-called 'MA-plot' to spot possible dye effects and other artifacts. The MA-plot is effectively a scatter plot of the intensities of the Cy3 and Cy5 channel that is rotated by 45°.

When plotting the normalized gene expressions of replicate arrays via multiple pairwise scatter plots, the points on these plots should scatter close to the line of equality on the arrays. Deviations from this line can indicate different types of problems. As with other methods, scatter plots on different scales can detect problems in different intensity ranges.

Correlograms

These are visualizations of the correlations between genes at increasing levels of separation on the array. They can indicate whether, and to what extent, bleeding has taken place. If correlations of genes near each other are not negligible, this can suggest that bleeding has taken place. They can also be useful in checking for normalization problems. However, care must be taken in their interpretation. This is particularly true if genes are not randomly laid out on the arrays in an experiment—here non-negligible correlations might simply reflect aspects of the layout of genes on the array.

5.3 Detecting Data Problems

Many things can go wrong when analyzing data from a complicated experiment. Here, four actual data problems are considered for their practical persistence and insidiousness—clerical errors, normalization problems, mishybridization and mishandling. A number of different methods are used to detect these different types of problems that can be found on a microarray. They are not the only methods that can be used nor are these problems the only ones that will be encountered. All methods used here are exploratory, in that they do not formally test for the problem-free state of the data. Assessing the quality of normalized data should be done with an open mind.

It is vital to know about the experimental details and the methods used. Or, at least, one should be able to ask those who do know. Only in this way can the cause of potential problems be found. This applies to the methods used for imaging and normalization, as well as to the actual physical experiment.

5.3.1   Clerical errors

The mislabelling of arrays or dyes is a very simple type of error. By mislabelling, we mean the administrative confusion of two or more arrays or dyes (for clarity, we do not take labelling, in this case, to mean the physical process of attaching dyes to the nucleotides). Most experimenters will have protocols that will aim to stop this error from happening. However, no matter how good a protocol is or how vigilant people are, mistakes will occasionally be made, and it is always worth checking for this. Further, other types of data-management errors can occur. For example, when processing the data, it is possible to wrongly align intensities to the genes. This could happen in Excel if one result is accidentally deleted and all the results in the column below are shifted up by one.

Although normalization should lead to arrays with comparable expression, this may not be the case when dyes have been mislabelled. If arrays have been mislabelled, it is the similarity of the patterns of expressions over each array that matters. Given this, it suffices to use correlation as a distance measure when checking whether arrays are mislabelled. If two dyes are accidentally swapped, the correlation between gene expressions may not be very indicative, whereas the mean distance might be. In such a case, a mean distance based measure is better to pick up such an error.

In checking for mislabelling, we can use visualization methods to see whether any arrays appear to be more similar to arrays associated with conditions other than their own.

Example: visualizing data handling problems with Sammon mapping. Here, we consider time-series gene expression data of *Mycobacterium tuberculosis* in a stressed growth stage from experiments conducted in 2000 at St. George's Hospital in London. The normalization used on the data follows the method applied by Wernisch et al. (2003). This normalization results in a single, normalized signal and reference intensity for each gene on each of the 16 different arrays used in this experiment. The log ratio of signal versus reference of each gene on each array constitutes the unit of comparison.

A two-dimensional Sammon mapping of these arrays, seen as high-dimensional vectors of log ratios, is given in Figure 5.4. It uses the distance measure $1 - \rho$ to calculate the stress between the arrays. Each array is referenced on the plot by the day the bacterium was 'harvested'—06 being on the sixth day of its growth, 14 being after 14 days of growth and so forth. The replicate number of the array is also given. For instance, the second replicate of the bacterium at 30 days is designated 30.2.

From this plot, it is clear that two of the arrays are noticeable outliers from the rest. The two most noticeable outliers are the first replicate of the day 14 results and the second replicate of the day 20 results (14.1 and 20.2 in the plot)—they are each particularly distant from their respective replicates.

Closer inspection of the normalized data revealed that in both these cases most of the gene's results for the signal had been shifted down one row so that they

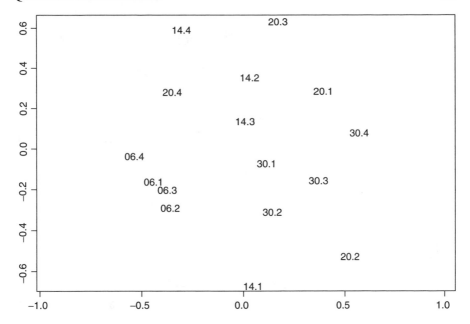

Figure 5.4   Sammon mapping of 16 arrays in tuberculosis data set using $1 - \rho$ distance. Four replicate arrays exist for each of four time points. The data used are the normalized log ratios of signal to reference.

represented the results for the neighbouring genes. The results for some genes from the fourth of the day 30 replicates (30.4 in Figure 5.4) had also been shifted. Far fewer genes were affected in this case, which is why the problem is not as noticeable as with the other two arrays.

Problems like these can occur extremely easily when handling and manipulating large data files. Visualizing the dissimilarities between arrays using methods such as Sammon mapping can help us spot these mistakes. Besides the Sammon mapping, simple pairwise scatter plots of replicate arrays can also provide insight into whether there have been administrative errors.

Example: array swap check with Sammon mapping.   The check of the arrays carried out above on the tuberculosis data found that a data handling problem had occurred with some of the arrays. Having rectified this, we can check the arrays again to see if any of the arrays have been swapped.

Figure 5.5 shows the two-dimensional Sammon mapping of the arrays created using the $1 - \rho$ distance measure. As with Figure 5.4, each array is referred to on the plot by the age the bacterium was 'harvested'—after 06, 14, 20 or 30 days of growth—and the replicate number of the array is given after this.

The figure suggests that it is not clear that there have been any array swaps. The fact that for over half the time points replicate arrays cluster together poorly is

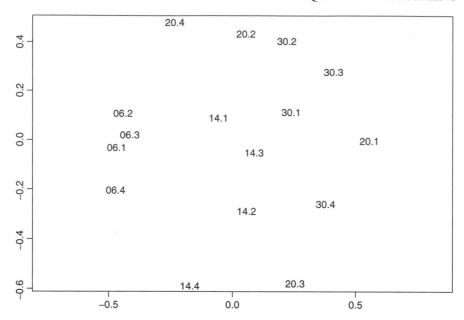

Figure 5.5   Sammon mapping of 16 arrays in tuberculosis data set using $1 - \rho$ distance. 4 replicate arrays exist for each of 4 time points (06, 14, 20 and 30 days).

not an indication of array swaps; in fact, this actually obscures our ability to tell if any of the arrays has been swapped. The poor clustering of the 14, 20 and 30 days replicates is more likely to be caused by some other problem with the experiment, such as poor hybridization or normalization (see Sections 5.3.3 and 5.3.2).

Mislabelling of arrays is most easily spotted when the experiments work well otherwise. If the arrays appear to cluster into distinct groups each corresponding to the different conditions with the exception of a pair of arrays from different conditions, which have switched clusters with each other, this can indicate array mislabelling.

A low-dimensional Sammon mapping can be calculated in R using the function sammon in the Venables and Ripley (1999) MASS library.

Example: dye mislabelling check with Sammon mapping.   Looking again at the tuberculosis data we can check for dye mislabelling. Figure 5.6 gives the two-dimensional Sammon representation of separate normalized signal and reference values for each of the four replicates taken at day six.

The day six array replicates were the most well clustered of all the time points (see Figure 5.5). That the arrays do replicate well can be seen from the fact that all of the signal arrays have negative values on the horizontal axis and all the reference signals have positive ones on this axis. Figure 5.6 gives no indication that any dye mislabelling has taken place.

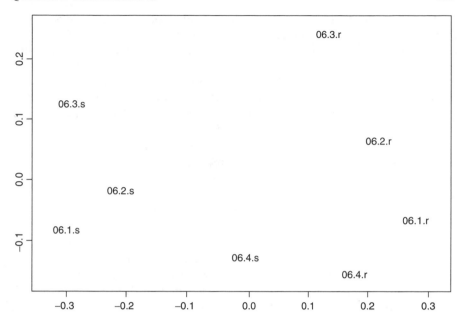

Figure 5.6   Sammon mapping of reference and signal values from the four replicate arrays for day six (06) in the tuberculosis data set using $1 - \rho$ distance.

The fourth replicate is the one array where the difference between the reference
and signal values is smallest. This replicate was the only one where the Cy3
dye was used for the reference sample rather than the signal sample. That this
array's signal and reference samples are so close suggests that the normalization
has not effectively dealt with the dye effect; this issue is considered further in
Section 5.3.2.

5.3.2   Normalization problems

Since normalization should aim to account for and remove a number of different
sources of variation, there are different ways of checking for the effectiveness of
normalizations. One way is to consider the relative similarity between arrays using
Sammon plots, as has been used for checking clerical problems. Another way is to
check for spatial patterns in images of the normalized intensities. Assuming that the
genes and controls have been randomly assigned to an array, a normalization should
display no spatial pattern—it should look like the picture seen on a traditional
television when there is no signal.

Because the normalized intensities can vary greatly in size, it is better to view
the ranks of these intensities as an image. This can highlight trends in arrays that
are difficult to spot otherwise, as the size of the largest intensities can impair visual
differentiation between smaller intensities.

Example: checking dye normalization of tuberculosis data using Sammon mapping. In the tuberculosis experiment, described above in Section 5.3.1, there are sixteen arrays with four replicate arrays at each of four time points. Although dye assignments were reversed in the experiment, this was done unevenly. Three out of the four replicate arrays from each time point had Cy3 dye applied to the treatment sample, whereas the reference sample was labelled with the Cy5 dye. For each time point, one of the four replicate arrays had the dyes applied the other way around—Cy3 for the reference and Cy5 for the treatment.

It is of interest to see whether this dye swap was fully accounted for by the normalization. Figure 5.7 gives a 2-D Sammon mapping plot of the data. Even if the intensities from two different dyes are highly correlated, this is still not useful if they are physically far apart. For this reason, the robust Manhattan distance (the distance d_1 of the family given in Equation (5.1)) was used, rather than the $1 - \rho$ dissimilarity measure.

The four arrays where Cy3 was used for the reference rather than for the signal are 06.4, 14.4, 20.3 and 30.4. From the plot it can be seen that all the arrays that have been applied with reference samples dyed using Cy3 rather than Cy5 are separated from the rest of their replicate arrays—the four dye swap arrays are all in the bottom right hand corner of the plot. This suggests that the dye effect has not been properly accounted for by the normalization.

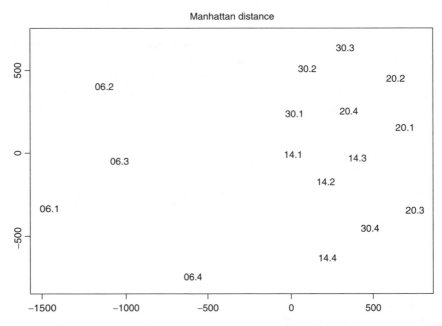

Figure 5.7   Sammon mapping of 16 arrays in tuberculosis data set using Manhattan distance. Four replicate arrays exist for each of 4 time points. The data used are the normalized log ratios of signal to reference.

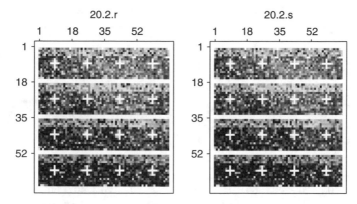

Figure 5.8   False arrays images on the rank scale of the normalized results of the second replicate from the 20-day old tuberculosis data. 20.2.r gives the reference normalized values; 20.2.s gives the signal normalized values. Note that the scale goes from white (low) to black (high).

Example: checking spatial normalization of tuberculosis data using 'false arrays plots'.   Many of the arrays from the tuberculosis data set show different patterns in their normalized values. An example is given in Figure 5.8 for the second replicate from the 20-day old bacterium condition, using the ranks of the normalized values of this array. There is an obvious trend for normalized values to decrease as we look from the bottom left to the top right of both the reference (20.2.r) and signal (20.2.s) images. The normalization used here does not account for spatial patterns, and this is picked up by considering such images.

A further point to note about these two images is that the first four rows of each of the four meta rows tend to have intensities lower than the other rows. This pattern is not due to a problem in the normalization but rather due to the design of the arrays, which has put genes and controls with low levels of expression in these rows. This pattern can be seen in all the arrays. Unfortunately, such deterministic layouts for genes and controls on arrays means it is very difficult to normalize the array properly—spatial patterns that naturally tend to occur across arrays cannot be accounted for easily.

5.3.3   Hybridization problems

Conducting microarray experiments is not straightforward. Sometimes, even with the best protocols in place, a whole array will fail to hybridize properly. A further problem can occur with two dye arrays—one channel may hybridize properly, but the other may not. These problems can sometimes be spotted at the image-analysis stage. However, they are not always so plain to see. Comparing replicate arrays against each other using visualization and clustering techniques can highlight poorly hybridized arrays. Because hybridization problems can affect arrays in a number

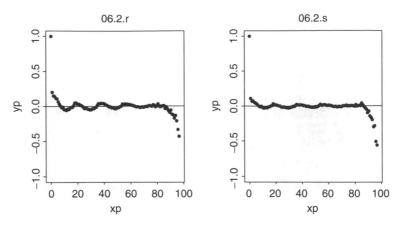

Figure 5.9   Correlograms of the normalized array images for a reference (06.2.r) and treatment (06.2.s) sample. The horizontal axes represent spot separation distances on each array; the vertical axes represent the correlations at these distances.

of different ways, we use mean distance–based measures to check for differences between arrays. As hybridization problems can affect anything from a small part of an array to the whole array, using a variety of mean-based distance measures is useful, such as using a variety of power distances (see Equation (5.1)).

Other problems with hybridization can involve so-called 'bleeding' or 'carry-over' effects, where the intensity of spots surrounding very high intensity spots is increased. If a microarray is well designed, with its genes and controls randomly assigned across the array, then this effect can be spotted by looking at the correlation between neighbouring genes. If this correlation is high, it would indicate the presence of spot bleeding.

Example: bleeding and non-random gene assignment.   Correlograms of normalized tuberculosis arrays can indicate whether bleeding effects have occurred. If no bleeding has occurred, then one would expect the correlation between neighbouring genes to fall off immediately from 1 to close to 0. With the tuberculosis data, however, a deterministic and highly organized design was used for the layout of genes and controls on each slide. It is therefore impossible to say whether bleeding is present, as from the design similar values for neighbouring genes are expected.

As can be seen from Figure 5.11, the first four rows of each meta row tend to have intensities lower than the other rows. This feature of the design manifests itself in the oscillating correlograms that these normalized arrays produce. An example of this is given in Figure 5.9 for the second replicate from the day 6 tuberculosis data. The high correlation between genes less than around 5 spots apart could be due to bleeding, but may equally be due to the fact that many of the comparisons between genes close to each other will be between genes that are both either in the first four rows of each meta row or in the remaining rows. The oscillation leads to peaks at distances of around 17, 34 and 51, which are multiples of the length of each meta row (17) and indicative that there is a confounding array design effect.

5.3.4   Array mishandling

Even under the strictest laboratory protocols, dust particles or a hair can get stuck onto a microarray. The array can become smudged or scratched when cover slips are placed on or removed from the slide. Damage can also occur while cleaning the slide after hybridization or in its storage. These problems seldom affect the whole slide, but they will affect at least some genes and we need to be aware of where they occur in order to account for these problems or to exclude the affected genes from the analysis.

These mishandling artifacts tend to have one of two effects on the genes they affect—the intensity will either be much lower than expected, or it will be much higher than expected. One of the best ways to spot these problems is by generating an image of the intensities. This is best done after normalization, as other effects that might hide these problems can often be removed by normalization.

To spot *high intensity* artifacts, looking at the intensities in their original scale is useful. Intensities in the original scale are generally positively skewed, with most intensities being considerably below the maximum intensity and a long tail of a few high intensities. Artifacts creating high intensities will be much easier to spot on this original scale. If the normalization transformed the intensities to a log scale, then re-exponentiating them before plotting is helpful.

To spot *low intensity* artifacts, it is better to transform the data to a symmetric or even slightly left-tailed distribution. An obvious choice is to transform the data from the original scale to the log scale. In this way, the artifacts stand out in comparison with most other genes.

As so often when considering microarray data, one should be aware that the genes may not have been assigned randomly to the arrays. If that is the case, the layout effect might confound with any mishandling problem.

Example: high intensity artifacts with tuberculosis data.   As an example of a high intensity artifact, the second replicate from the 14-day old bacterium is shown in Figure 5.10. This shows how at the bottom of both the reference and signal images the intensities tend to be much higher than over the rest of the array. An artifact this large should normally be dealt with by removing spatial trends in the normalization process. However, other artifacts may be too small to be highlighted using normalization.

Example: low intensity artifacts with tuberculosis data.   One of the arrays from the tuberculosis data set has a low intensity artifact that is very clearly

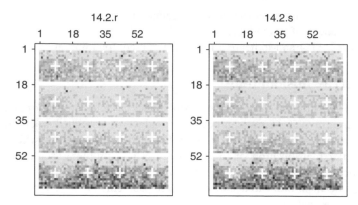

Figure 5.10   False array images of the normalized results on the original scale of the second replicate from the 14-day old tuberculosis data. 14.2.r gives the reference normalized values; 14.2.s gives the signal normalized values. Note that the scale goes from white (low) to black (high).

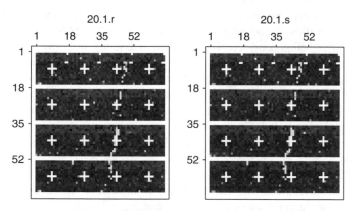

Figure 5.11   False array images on the log scale; 20.1.r and 20.1.s give the reference and treatment values of the first day 20 replicate in tuberculosis data. The scale goes from white (low) to black (high).

visible on the log scale. Figure 5.11 shows, on the log scale, both the reference and treatment normalized values for the first replicate of tuberculosis bacterium harvested at day 20. The artifact appears to be a hair or perhaps a scratch, going across all the rows of the array in a line starting near column 52 and ending near column 35.

Note that displaying either of these images on their original scale does not allow this artifact to be seen—(see Figure 5.12).

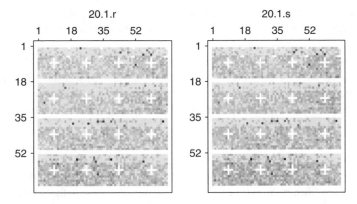

Figure 5.12   False array images on their original scale of the normalized results of the first replicate from the 20-day old tuberculosis data. 20.1.r gives the reference normalized values; 20.1.s gives the signal normalized values. Note: the scale goes from white (low) to black (high).

5.4   Consequences of Quality Assessment Checks

Having done checks of the normalized data and found problems, some further course of action is required. This may be as simple as altering the labelling when finding that an array has been mislabelled; however, it may be more complicated and may involve a decision about whether, and how much, data should be removed from the analysis.

Although throwing out any arrays where there are real problems is safe, it can also be very wasteful. It may be that the problem is localized to one part of the array and only the affected spots should be removed; or else, it may be that redoing the normalization, with, for example, a spatial effect included, resolves the problem. Moreover, sometimes what may appear to be a problem with the data turns out in fact to be natural variation in the data. In this case, throwing out data might in fact introduce bias. One should really convince oneself that there is a serious problem before omitting data. This possibility should always be checked.

Clerical errors.   If a clerical problem is spotted, then it is typically trivial to sort this out; one can go back, make sure that the data are associated with the correct gene labels and then carry on. Obviously, data normalization should be done again on the correctly labelled data.

Normalization.   If there appears to be a problem with the normalization, then it may be possible to change the normalization method to account for this. If the dye effect has not been accounted for properly, then it should be possible to change the normalization to account for this; spatial trend or other systematic effects may be tackled in a similar fashion.

If, after trying different normalizations, the problem appears to remain, then it may be necessary to remove the offending array from the analysis.

Mishybridization.   If a whole array has mishybridized, then the data gathered from it will be useless. However, careful checking that this is the case should be done before getting rid of any arrays from the analysis. Check that the apparent problem has not been caused by something other than mishybridization, including natural variability.

Mishandling.   If the problem is a local one, then the spots affected may have to be removed. If it affects the whole array, then the whole array should be removed. Neither of these things should be done as a matter of course—the data needs to be checked for other possible causes of the unusual observations.

6

Microarray Myths: Data

Several 'myths' about microarray design, data handling and normalization have entered the microarray lexicon. The aim of this chapter is to discuss some of these myths and, in some cases, expose their absurdities. The chapter is organized as a loose collection of sections, each dealing with a different myth.

6.1 Design

6.1.1 Single-versus dual-channel designs?

Design considerations for two-channel arrays have received far more attention than the design of its single-channel cousin. The claim seems to be that the design for a single-channel array is much simpler. Within-block comparisons of conditions are impossible for single-channel arrays, and there are no differential dye efficiencies that might complicate comparisons in other ways. On the other hand, if we regard each channel on a two-channel microarray as an independent set of observations, then there is no real difference between a two-channel and one-channel microarray design. But can we really make this assumption? This section discusses the advantages and disadvantages of considering a two-channel microarray simply as a collection of two one-channel chips.

Thesis: '1 two-channel array ≠ 2 one-channel arrays'

The earliest microarray designs compared either two conditions of interest directly on a single slide or one condition of interest versus some meaningless, administrative reference. In either case, no information was lost considering only the log ratios of the green and the red channel of each spot.

An argument why a pair of observations on a single spot cannot be considered as independent observations is because they share the same nuisance effects. They

Statistics for Microarrays: Design, Analysis and Inference E. Wit and J. McClure
© 2004 John Wiley & Sons, Ltd ISBN: 0-470-84993-2

are likely to have a high level of spatial correlation, and obviously they share the same print-tip effect. Moreover, each of the two samples compete for the same hybridization positions on the slide. For highly expressed genes, that is, genes that reach signal saturation levels, this may introduce negative correlation. In Figure 6.1, the channels of four two-channel microarrays have been scaled into a two-dimensional plane. It shows how the channels are closer to one another than they are to their biological replicates. The cumulation of nuisance effects seems at least as large as the contribution of any biological variation.

This observation alone shows clearly that treating the spot pair as independent observations is unsatisfactory. One possibility is to consider the data as a bivariate vector and model them explicitly as such (Glasbey and Ghazal 2003). More commonly, the observation pair is summarized into a single log ratio of the red versus the green channel. This approach has a significant impact on the design. As a result of this approach, the only way to compare expressions from two conditions in a dual-channel array experiment is by linking a chain of microarray experiments from one to the other condition (Yang and Speed 2003).

Let y_i be the log expression of a particular gene subject to condition i. Assume that an experiment consists of two microarrays and three conditions; conditions 1 and 3 are hybridized to the first array and conditions 2 and 3 are hybridized to the

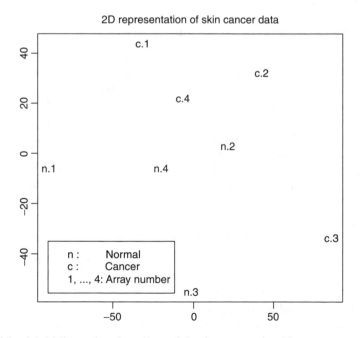

Figure 6.1   Multidimensional scaling of 4 microarrays in skin cancer example. It can be seen that expressions on the same array are more similar to each other than to their respective replicates.

second array. In order to compare the gene expression under condition 1 with that under condition 2, one would consider the difference between two log ratios,

$$\left(y_1 - y_3^{(1)}\right) - \left(y_2 - y_3^{(2)}\right).$$

Figure 6.2 illustrates this fact.

Antithesis: '1 two-channel array = 2 one-channel arrays'

There is another side to the coin. Although there are good reasons to treat a green–red observation pair differently from two independent observations, the opposing camp might claim that it has been mainly inertia that has kept biologists considering only this option. There are other good reasons to use each of the channels as individual observations.

For example, the variance of a log ratio consists approximately of the sum of the variance of each of the log expressions. In short, whereas the log-ratio approach might give some protection against bias, the direct approach tends to have smaller variances.

The stark contrast between the analyses of one-channel and two-channel arrays begs the question as to what is the precise nature of the difference between the two chip types. We focus on the individual probe level. A simplistic model for the expression of a single probe for a particular condition c on a one-channel array includes a possible spot effect and an array effect:

$$y_{csa} = \mu_c + S_{as} + A_a + \epsilon_{csa},$$

whereas for a dual-channel array this also involves a dye effect:

$$y_{csad} = \mu_c + S_{as} + A_a + D_d + \epsilon_{csad}.$$

We assume that the nuisance effects are systematic, fixed effects, except perhaps for an additive random component. Standard practice for measuring differential expression in a one-channel array experiment focuses on the difference of two (or more) expressions, $y_{111} - y_{222}$. This comparison has an associated imprecision, measured in terms of bias and variance,

$$\text{Bias}(y_{111} - y_{222}) = E\left((y_{111} - y_{222}) - (\mu_1 - \mu_2)\right)$$

$$= A_1 - A_2 + S_{11} - S_{22}$$

$$V(y_{111} - y_{222}) = 2\sigma_\epsilon^2.$$

In single-channel arrays, each comparison may be fraught with bias due to unequal array 'brightness' and unequal spot size. Proper normalization might be able to deal with the array bias but not with the spot bias. The importance of this effect depends on the physical reproducibility of the probe spotter. Averaging the probes into a single gene effect per slide might diminish the spot effect.

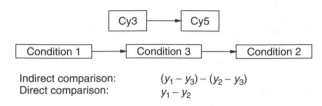

Figure 6.2   Possible direct and indirect comparisons of conditions 1 and 2 when performing a dual-channel chip experiment involving two arrays.

What has been said for a single-channel microarray experiment could translate immediately into a comparison strategy for a two-channel microarray experiment. Rather than using indirect comparisons via array-specific log ratios, one could simply use all available channel data, ignoring the fact that they came from the same or different arrays. These two philosophies are summarized in Figure 6.2.

The difference between these types of comparisons can be evaluated by considering the bias and variance of these comparisons. For the *direct comparisons* there is a bias and a small variance,

$$E\left((y_{1;111} - y_{2;222}) - (\mu_1 - \mu_2)\right) = A_1 - A_2 + S_{11} - S_{22} + D_1 - D_2$$

$$V(y_{1;111} - y_{2;222}) = 2\sigma_\epsilon^2,$$

whereas for the *indirect comparisons* there is no bias but a larger variance,

$$E\left(((y_{1;111} - y_{3;112}) - (y_{3;221} - y_{2;222})) - (\mu_1 - \mu_2)\right) = 0$$

$$V\left((y_{1;111} - y_{3;112}) - (y_{3;221} - y_{2;222})\right) = 4\sigma_\epsilon^2. \qquad (6.1)$$

Although the direct approach is not without dangers, if it is combined with effective normalization and the spot is highly reproducible, it can be superior to the approach with log ratios.

Synthesis

As always, the truth lies somewhere in the middle. Many of the nuisance effects can be suitably eliminated by effective normalization methods, such as those that we propose in Chapter 4. If the deterministic nuisance effects are removed, then a lot of the correlation between the two channels will disappear. This would make a case for using the two channels of a dual-channel array as two one-channel chips.

On the other hand, several authors have suggested that spot effects are an issue for dual-channel arrays (Kerr et al. 2000). If these effects are purely individual and do not correlate with, for instance, pin or spatial effects, then these spot effects cannot be eliminated by normalization without sacrificing half of all degrees of freedom. Such a normalization would be equivalent to taking log ratios. Moreover, although normalization might account for the *deterministic* components of the nuisance effect, it cannot eliminate the correlation in the random error. This means that the variance of a log-ratio comparison, such as in Equation (6.1), might not be as large as $4\sigma^2$.

Which of the two comparison strategies is better depends ultimately on the relative bias effect as compared to the variance increase. In principle, bias is an undesirable effect because it is impossible to detect it by only measuring variation. Perhaps, with the advance of technology, spot effects might become less important. Eventually, which approach is better is an empirical and not a theoretical question. We recommend that the design of a two-channel chip experiment should start with the assumption that it will be using the log ratios. The optimal design considerations in Section 3.5.1 will aid the biologist in this choice. After the experiment has been performed, one can always decide later to use both channels individually.

6.1.2   Dye-swap experiments

Dye-swap experiments have been recommended for averaging out the dye effects in the two-channel arrays. There are two ways the data from a pair of dye-swapped arrays can be used. Either the data can be averaged before continuing, with only the averaged data being used for further analysis, or the data from both arrays can be used. In the first case, an enormous price is paid for a rather simple dye normalization. Twice the number of arrays are needed to achieve a dye normalization. This seems a waste of resources. However, the other option is not without problems either.

Even after careful normalization, two different conditions on the same array look frequently more similar than the same condition on two different arrays. Performing a dye-swap experiment means that twice a pair of conditions ends up having similar expression values, not due to any biological similarity, but simply because they appear twice on the same array together. Especially, in large studies where the number of conditions exceeds the possibility of studying every possible pair, choosing a dye-swap design will increase the amount of experimental bias in the study.

Using a dye-balanced design instead of dye-swap design has many of the advantages of dye swap without any of its disadvantages. See Section 3.3.2 for more details.

6.2   Normalization

6.2.1   Myth: 'microarray data is Gaussian'

The theory of statistics is largely built on the idea that data have some component of randomness, that is, follow some probability distribution. This means that if the same experiment were repeated many times, the observations would cluster in a particular way around some kind of *true* expression of interest. By far, the most

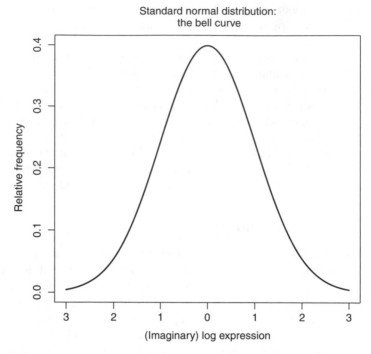

Figure 6.3   Probability density function for a standard normal distribution.

common distribution is the *normal* or *Gaussian* distribution. It is characterized by its bell shaped curve, such as in Figure 6.3.

If the distribution were normal, then a lot of theory would be available, which would be a great relief to the bioinformatics practitioner. It is not surprising, therefore, that a lot of attempts have been made to show that log-expression data from microarray experiments are normally distributed. It is a common misunderstanding that the large amount of spots on the microarray guarantee that the data on the microarray should be normally distributed. In order to reinforce this point, a quantile–quantile plot (QQ-plot) of the data in a single array, such as that shown in Figure 6.4, is sometimes presented. The fact that the QQ-plot is close to a straight line would, so the argument goes, show that the logarithms of the expression data are indeed normally distributed.

However, there is a fundamental flaw in this argument. The distribution we are interested in from the point of view of inference is *not* the distribution of the gene expressions over the array but the distribution of a single gene expression across many arrays. Remember that a probability distribution is (operationally) defined as a histogram of the repetitions of the *same* experiment. Looking at the QQ-plot for many different genes on the same array gives an idea of the distribution of the expression levels of the genes in the experiment but not of a *probability* distribution.

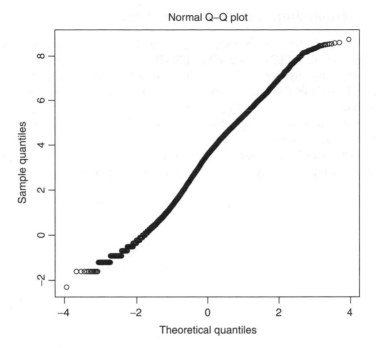

Figure 6.4   Quantile–quantile plot for log-expression values in array 27 of the mammary gland experiment.

6.2.2   Myth: 'microarray data is not Gaussian'

The QQ-plot argument that the data are normally distributed can easily be turned against itself if one looks at another set of data. Yang and Speed (2003), for example, report a particular QQ-plot that deviates quite substantially from the normal distribution. They consider t-statistics for all genes in the microarray experiment. If the argument were to be made that this shows that microarray gene expression data are *not* normally distributed, then this will also be a fallacy. Roughly speaking, probability distributions can only be assigned to repetitive measurements of different instances of the same process. The collection of expression summaries for different genes is a distribution but not a probability distribution.

The real question of interest is whether for each gene the distribution of expression *across* all the arrays is normal. The strictly positive nature of gene expression values makes a normal distribution an unlikely candidate. However, after a logarithmic transformation of the data, it is our experience that many gene expressions are indistinguishable from a normal distribution. We certainly do not claim that log expressions are normally distributed *per se*. The true underlying distribution is probably more complicated than that. However, it is our experience that the misspecification made by using a normal approximation is typically negligible.

6.2.3 Confounding spatial and dye effect

A lot of normalization techniques are presented as a sequential toolbox (e.g. Wolkenhauer et al. (2002); Yang et al. (2002b)). Does it matter in which order the normalizations are applied? The answer is a yes. In this section, we show that normalizing the dye effect *before* the spatial effect can lead to disastrous consequences.

A common way to deal with the dye effect is by means of a smoother normalization. First, a smoother is applied to the scatter plot of the expressions in both channels. The idea is that the smoothed line represents pairs of 'equivalent expressions' in the two arrays. Deviations with respect to this line characterize the differential expression of that EST for those two tissues. Figure 6.5 shows the log-transformed raw data from both channels of one of the arrays from the skin cancer

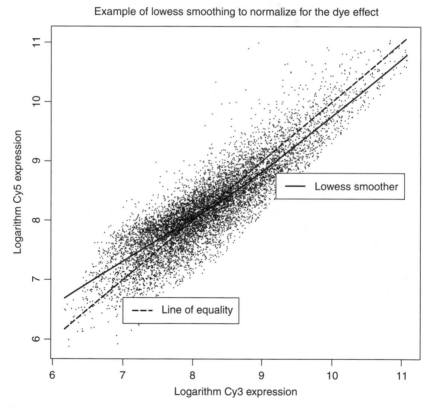

Figure 6.5   Raw log-transformed data from a cDNA slide from the skin cancer experiment. The *x*-axes and *y*-axes contain the Cy3 and Cy5 values respectively. The lines in both plots correspond with the line of equality and a lowess smoother through the points.

experiment. If no dye bias was present and most of the genes were not differentially expressed, then we expect the majority of the genes to be expressed around the line of equality. This is clearly not the case. The lowess smoother shows how, instead, the data lie on a slightly curved line. Lowess normalization is performed effectively by bending the smoothed line to the line of equality and thereby taking the points along. Further details can be found in Section 4.3.4.

However, one has to be careful performing this type of normalization as there can be many confounding factors that will undermine this normalization. Consider Figure 6.6 of an array from the same skin cancer experiment.

6.2.4   Myth: 'non-negative background subtraction'

The presence of background signal in microarray images is beyond dispute. Some dye appears to adhere to parts of the array where there should not be any signal. The strength of this background signal appears to vary across the array and to affect within- and without-spot pixel intensities equally. This non-specific hybridization on the spots is problematic, and we would like to remove it.

The most common approach to background correction is to try to subtract an estimate of the background from the signal. This estimate is often calculated for

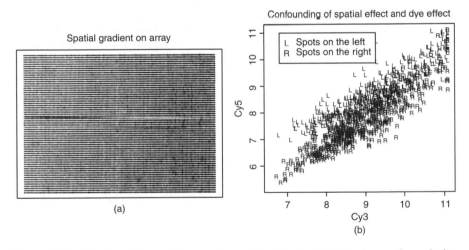

Figure 6.6   Confounding of dye and spatial effect: plot (a) shows the relative distribution of the log-expression ratios over the third skin cancer array. It is clear that the log ratios vary in size as a function of the position on the array. The further to the right, the larger the log ratios tend to become. When plotting the log Cy3 expression versus the log Cy5 expression in (b), a curious effect seems to occur. Rather than the typical 'banana' shape, it looks as if there are several groups, each with their own dye effect. It turns out that the top group corresponds to the genes on the left of the array, whereas the lower group contains the genes on the right of the array. Values in the centre are located in the centre of the array.

each spot from the pixel intensities in the area immediately surrounding that spot. One perceived problem with this method, and many similar background subtraction methods, is that the intensities from some of the spots on the array will be *less* than their associated background intensities. This leads to *negative estimates of the background corrected intensity*. This seems undesirable since it is impossible for a gene to produce a negative amount of RNA.

For this reason, much effort has been spent or potentially wasted in trying to ensure that background subtraction methods do not give negative results. The potential is that the desire to create such methods can lead to either very poor background correction procedures, which increase the noise in the data, or to meaningless procedures. For example, Affymetrix (2001) suggests an intricate way of subtracting a small signal quantile from the observed signal and replacing any negative values by some small quantity. This method does not seem to be based on any biochemistry nor does it lead to any noticeable data quality improvements.

Two points should be noted about background subtraction methods. The first is that both the estimate of the spot's intensity and its *local* background will inevitably be measured with error; subtracting one noisy estimate from another increases the variability. Secondly, although negative estimates of gene expression are wrong, there is little reason why negative values cannot be set to zero.

Section 4.1.1 shows how a carefully tuned and robust local estimate of the background using a top-hat filter can eliminate background *and* reduce spot variability. Other sensible choices for stable estimates are global background estimates, such as the global average background intensity μ_e used in Equation (4.2). Section 4.3.3 gives more detail on both global and local background correction methods for two-channel arrays. Within Section 4.4.2 there is information on Affymetrix background correction procedures.

Part II

Getting Good Answers

7

Microarray Discoveries

Microarrays are, foremost, tools of discovery. They are good at unearthing small gems of knowledge from the wasteland of ignorance. Two main requirements to make the most out of microarray technology are (i) the ability to search large and high-dimensional data sets in a transparent and fast manner and (ii) the ability to represent and visualize patterns in these data sets in order to communicate with the biologist.

This chapter is a gene-expression adventure. It aims to give an overview of pattern finding and visualization or 'communication' techniques. When applying the methods in this chapter, it is essential to have a biologist nearby. Few things are worse than answering a question that was not being asked. Moreover, keen biological insights are necessary for the interpretation of many results.

There are wonderful opportunities to combine the exploratory capacity of microarrays with the knowledge of medical researchers, geneticists and other biologists. It is an exciting time.

7.1 Discovering Sample Classes

When several tissue samples of a certain type are hybridized to a microarray in an experiment, it is natural to ask whether each of these tissues can be grouped in homogeneous subtypes. For example, it is well known that cancer is a term that covers many different genotypical configurations, even if they have the same phenotype. Several experiments involving cancer biopsies have been aimed at finding cancer subclasses (Golub et al. 1999). To the extent that these subclasses are the result of differential *expression* of the genes, clustering of the gene-expression profiles may reveal these classes.

In much of the DNA sequence analysis, the aim was to find phenotypes that were associated with a particular genomic DNA pattern. For example, severe combined immune deficiency (SCID), which incapacitates an immune response

Statistics for Microarrays: Design, Analysis and Inference   E. Wit and J. McClure
© 2004 John Wiley & Sons, Ltd   ISBN: 0-470-84993-2

in children, is caused by a single base-pair change in a single gene, namely the *ADA* gene. With the advent of microarray technology, it is also possible to detect other types of genetic peculiarities, for instance, particular amplification and deletion patterns of a part of some genomic DNA. Repetition patterns of DNA can have a crucial impact. The gene that causes Huntington's chorea is a repetition of a single word: CAG, CAG, and so on. If the number of repeats is more than 39, then the person will develop this debilitating disease; otherwise he or she is fine. Microarray experiments in which the hybridization samples consist of genomic DNA might also detect sequence peculiarities of this latter type. In the breast cancer example (Section 1.2.3), the values from the microarray do not represent gene expression values but quantifications of the number of genomic copies of a particular gene. In this way, amplification and deletion patterns of particular sequences can be monitored and possibly correlated to certain clinical outcomes.

7.1.1   Why cluster samples?

Although there are very few formal theories about clustering, the two main intuitive ideas of what constitutes a cluster are internal cohesion and external isolation. The reason why this might be interesting depends on the type of microarray experiment considered.

Time-course experiments sample a particular type of organism at different stages in some development. Clustering here emphasizes developmental similarities. For example, a couple of weeks after a mouse has finished lactating, the gene expressions of its mammary gland look very much like those in a mature virgin mouse (*cf.* Figure 7.1). Besides clustering, visualization of the individual slides through multidimensional scaling can reveal very similar things.

Comparative experiments compare the effect of several conditions on the gene expressions. These conditions can be set up as a one-way, two-way or any higher-order experiment. Closely related experiments might test the effect of a certain amount of dose of some medicine on the transcription of the genes. In all of these experiments, the sampled individuals are (considered) homogenous replicates of one or more populations of interest. Clustering and multidimensional scaling in these settings are mainly of interest for quality control purposes. If an observation does not cluster with its replicates, it should be decided whether this is part of natural variation, and therefore should be left in the sample, or whether this is the result of poor normalization or a failed hybridization, in which case other actions might be required.

Clustering is most appropriate in typical clinical experiments. In these experiments, the experimental units are *a priori* considered heterogeneous samples corresponding to one or more phenotypes. Even if the samples come from the same phenotype, it could be of interest to investigate if there are several *distinct* genotypes corresponding to it. Clustering might be able to reveal these subtypes and suggest differential treatment as a result.

There is one more reason why clustering may be useful in all of these cases, although this relates more closely to the material of the following chapter. By

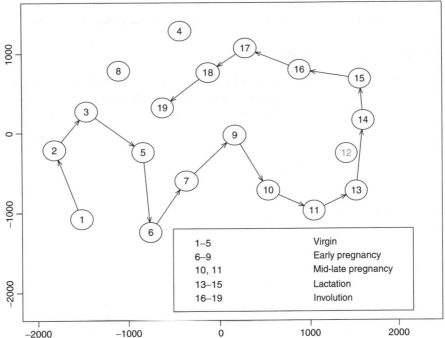

Figure 7.1   Clustering algorithms can be used to bring to the fore biologically important information. Clustering the mammary gland tissues clearly shows the temporary development of the overall gene expression. The lactation phase is particularly distinct from the other phases.

subdividing the samples in distinct homogeneous groups of individuals, it might be possible to relate their gene expressions to a particular response of interest separately. This is closely related to the so-called *k-nearest neighbour* classification method, discussed in Section 9.3.3.

7.1.2   Sample dissimilarity measures

Whether two objects are considered 'close' depends on the particular quantity of interest that is being compared. What is it that is required for the expression snapshot of different samples to be similar? An expression snapshot is similar to the next when the values are actually close. There could be circumstances in which highly correlated expressions are considered similar, especially if all the expression profiles have not been accurately normalized. Churchill (2002) and others have found correlation useful to compare sample similarities in the context of design (see also Section 3.2.3). The correlation coefficient is on a fixed scale and can easily be interpreted. Our quantile normalization (Section 4.3.5) guarantees that all arrays are approximately on the same scale. If this is indeed the case, it should not be necessary to consider correlation-type similarities, which are particularly suitable when comparing variables on different scales, for clustering samples.

Geometric distance measures

When you consider the samples as points in a high-dimensional gene expression space, it is natural to use a geometric measure to evaluate the distance. These measures are the so-called L_p norms. The most common of these are the Manhattan distance or L_1,

$$d_1(x, y) = \sum_{i=1}^{n} |x_i - y_i|,$$

and the Euclidean or L_2 norm,

$$d_2(x, y) = \sqrt{\sum_{i=1}^{n}(x_i - y_i)^2}.$$

The larger the value of p, the more sensitive the L_p measure is to a few large gene expression differences. The Manhattan distance is the most robust of all the geometric distance measures and least influenced by outlying gene expressions.

Chipman et al. (2003) suggest scaling the genes before calculating the L_p distance, so that they have mean zero and variance one. This follows the general recommendations of some of the same authors in Hastie et al. (2001) for variables of different types. Since the data involve gene expressions for each of the genes, we believe that centering and scaling is not absolutely necessary and might in fact introduce spurious (dis)similarities. However, the variance for highly expressed genes is typically higher, and we do recommend to perform some variance stabilizing transformation, such as the logarithm, before calculating sample similarities. Other variance stabilizing functions that take into account additive noise in the lower channels, such as those suggested by Rocke and Durbin (2001) and Huber et al. (2003), are also possible.

Weighted measures

The geometric L_p norms calculate the distance between two samples as the sum of individual gene contributions. In general, when comparing two samples, it is possible to weight the contribution of each gene by a measure of confidence for that particular gene. A measure of accuracy is typically available in terms of the standard error of each spot. Highly variable spots might produce a spurious large difference between two samples. By down-weighting such spots, a more stable distance measure results. This idea was used in a microarray context by Hughes et al. (2000).

If we think of each component $|x_i - y_i|^p$ as an estimate of the true distance between samples x and y, then these estimates are typically combined optimally by weighting them by the precision. For the gene expression x_i for gene i, the precision is typically available through the spot variation of the pixels that make up that spot. The precision of the spot means x_i and y_i are defined as the inverse of the square of the standard errors. The standard error of x_i is $se_{xi} = s_i/\sqrt{n_i}$, where s_i is the pixel standard deviation of x_i and n_i is the number of pixels in the spot.

We redefine the weighted distance between two samples x and y according to the L_p norm as,

$$d_p^w(x, y) = \sqrt[p]{\sum_{i=1}^{n} w_i |x_i - y_i|^p},$$

where

$$w_i = \frac{(se_{xi}^2 + se_{yi}^2)^{-1}}{\sum_{j=1}^{n}(se_{xj}^2 + se_{yj}^2)^{-1}} \, n. \tag{7.1}$$

The weights add up to n, analogous to the unweighted case. The weights for gene i are proportional to the inverse of the sum of the standard errors of the two corresponding gene expressions x_i and y_i. Typically, the values $p = 1$ or $p = 2$ are used corresponding to the weighted Manhattan and weighted Euclidean distance respectively. If other measures of precision are available rather than the spot standard errors, then they can be substituted for se_{xi} and se_{xi} in Equation 7.1.

The previous weights are not optimal for $p > 1$, since they are not proportional to the precision of $(x_i - y_i)^p$ but only to the precision of $(x_i - y_i)$. However, the 'improvement' using a first-order approximation,

$$w_i = \frac{[(x_i - y_i)^{2p-2}(se_{xi}^2 + se_{yi}^2)]^{-1}}{\sum_{j=1}^{n}[(x_j - y_j)^{2p-2}(se_{xj}^2 + se_{yj}^2)]^{-1}} \, n.$$

is itself highly variable and is not recommended. We suggest that the reader sticks to the (suboptimal) choice of weights in Equation 7.1.

Distances between clusters of observations

Several clustering algorithms rely on their ability to define the distance between clusters of observations. For example, in hierarchical clustering, one considers whether two groups of observations are close enough in order to be put together. But what does 'close' really mean when you do not compare two unique geometric points?

There is no single answer to this question. It depends on two decisions the observer has to make in order to be able to measure the distance between groups of points: (1) **distance:** the original distance measure used to measure the distance between two points (2) **linkage:** condensation of each group of observations into a single representative point. There are many ways in which the second choice of linkage can be implemented. We restrict ourselves to discussing those that are standardly implemented in R.

- **Average linkage.** The distance between two groups of points is the average of all pairwise distances.

- **Median linkage.** The distance between two groups of points is the median of all pairwise distances.

- **Centroid method.** The distance between two groups of points is the distance between the centroids of both groups.

- **Single linkage.** The distance between two groups of points is the smallest of all pairwise distances.

- **Complete linkage.** The distance between two groups of points is the largest of all pairwise distances.

Single linkage and complete linkage represent the two opposite ends of the cluster distance spectrum. In practical clustering problems, without much *a priori* information they are not particularly useful. The other types of group distances lie somewhere in between and have a comparable performance. Average linkage is typically used in hierarchical clustering approaches, whereas a variation of the centroid measure is employed for multidimensional scaling of clusters in the family of partitioning-around-medoids (PAM) methods (Kaufman and Rousseeuw 1990), discussed later.

Measures for repeated observations

Many microarray experiments contain technical replicates (see Section 3.2). These arrays are naturally close to one another, and clustering them as individual arrays may only confirm that they are close. Whereas this can be interesting from a quality control point of view, from the point of view of inference this does not add any information from what we knew already, namely, that they are replicates. One way of using the replicate arrays is by taking the spot-wise average of every condition before clustering. This leads to more stable gene expression values for each condition. Normal Euclidean or Manhattan distance measures can be used in subsequent clustering of the condition averages. This is, however, not necessarily the most optimal use of replicate observations.

If several samples are measured on more than one array, there is the ability of testing the stability of the clustering using the repeated observations. Similarly, if a particular condition of interest is hybridized to several arrays, the gene expression variability of that condition can be taken into account explicitly when clustering that condition. There are several ways of using these repeated measures. Kerr and Churchill (2001b) use bootstrapping to assess *post hoc* the stability of a cluster. Yeung et al. (2003) make explicit use of the replications to improve cluster stability. In this section, we pursue two ideas to use the replications to adjust the dissimilarity matrix.

Using replicates for calculating precision weights.   If replicates are available for the conditions, another confidence measure is available for each of the genes. In particular, the distance between the mean vectors $\bar{x}$ and $\bar{y}$ can be weighted by the repeatability of each of the gene expressions across the replicates. This suggests that we can adapt the weighted L_p norms, defined above, to define a new distance measure.

Let x_{ij} and y_{ij} stand for the jth replicate of the expression of gene i under conditions $\mathcal{X}$ and $\mathcal{Y}$ respectively. We calculated $\bar{x}_i$ and $\bar{y}_i$, the average gene expression, as well as se_{xi} and se_{yi}, the sample standard errors under conditions $\mathcal{X}$ and $\mathcal{Y}$ respectively. The weighted L_p distance between conditions $\mathcal{X}$ and $\mathcal{Y}$ is defined as before,

$$d_p^w(\mathcal{X}, \mathcal{Y}) = \sqrt[p]{\sum_{i=1}^{n} w_i |\bar{x}_i - \bar{y}_i|^p},$$

where

$$w_i = \frac{(se_{xi}^2 + se_{yi}^2)^{-1}}{\sum_{j=1}^{n}(se_{xj}^2 + se_{yj}^2)^{-1}} \, n.$$

Comparing distributions.   Whenever there are replicate arrays for several conditions, it is natural to compare the expression *distributions* between these conditions, rather than the individual arrays. A natural candidate for a distance measure between two distributions is the *Kullback–Leibler* (KL) divergence. For two distribution F and G with densities f and g respectively, the KL-divergence between the two distributions is defined as

$$d_{\mathrm{KL}}(F, G) = \int_{-\infty}^{\infty} f(x) \, \log \frac{f(x)}{g(x)} \, dx.$$

However, the KL-divergence is not a distance measure, as it lacks symmetry. The KL-divergence from G to F is not necessarily the same as from F to G. This wracks havoc in distance-based clustering algorithms. There have been several suggestions on how to symmetrize the KL-divergence, such as

$$d_{\mathrm{Jeffreys}}(F, G) = \frac{d_{\mathrm{KL}}(F, G) + d_{\mathrm{KL}}(G, F)}{2},$$

which corresponds to a proposal from Jeffreys (1946). Johnson and Sinanović (2001) suggest that the harmonic mean,

$$\frac{1}{d_{\mathrm{Resistor}}(F, G)} = \frac{1}{d_{\mathrm{KL}}(F, G)} + \frac{1}{d_{\mathrm{KL}}(G, F)},$$

which they name the *resistor-average* distance, has superior geometric properties.

Another difficulty is determining the correct distribution for the high-dimensional gene expression data. Especially, with few replicates it is impossible to fit any complicated distribution. From a computational point of view, the assumption of a multivariate normal with a diagonal variance–covariance matrix is convenient, and we have found it quite useful in practice. Correlations between the expressions of certain genes certainly exist, but we have generally too few degrees of freedom to even start to estimate them. If r_F and r_G replicate arrays are spotted for conditions F and G respectively, p is the number of genes, and we indicate with

$$f_{ij} : i = 1, \ldots, p, \quad j = 1, \ldots, r_F$$

$$g_{ij} : i = 1, \ldots, p, \quad j = 1, \ldots, r_G$$

the gene expressions for the two conditions, then the empirical Kullback–Leibler divergence for the two conditions under the assumption of normality can be calculated as

$$\hat{d}_{\mathrm{KL}}(F, G) = \sum_{i=1}^{p} \left[\log \frac{\hat{\sigma}_{gi}}{\hat{\sigma}_{fi}} + \frac{1}{2} \left(\frac{\hat{\sigma}_{fi}^2}{\hat{\sigma}_{gi}^2} + \frac{(\hat{\mu}_{gi} - \hat{\mu}_{fi})^2}{\hat{\sigma}_{gi}^2} - 1 \right) \right],$$

where

$$\hat{\mu}_{fi} = \frac{1}{r_F} \sum_{j=1}^{r_F} f_{ij},$$

$$\hat{\sigma}_{fi}^2 = \frac{1}{r_F - 1} \sum_{j=1}^{r_F} (f_{ij} - \hat{\mu}_{fi})^2$$

are the sample mean and sample variance of the observations associated with the ith gene. With this empirical KL-divergence, the empirical versions of the two distance measures d_{Jeffreys} and d_{Resistor} are easily calculated.

7.1.3 Clustering methods for samples

Clustering is the name for an ever-increasing set of techniques that aim to divide the observations into homogeneous subsets. In the case of gene expression data, these subsets of RNA samples are candidates for classes of distinct genotypes that perhaps can be correlated with other data obtained on these samples.

The most commonly used clustering method in microarray studies is hierarchical clustering, introduced in the field by Eisen et al. (1998). This method has an appealing advantage in that it produces a linear ordering of the objects that are being clustered. Visualization is an important aspect for any clustering method, particularly in the case of high-dimensional data. By being able to arrange clusters with respect to one another visually, important insights might be obtained that otherwise would remain hidden. For example, the multidimensional scaling of the arrays in the mammary gland example (see Figure 7.1) shows clearly in a two-dimensional representation of the 12,000 dimensional data how involution of the mammary gland after pregnancy and lactation genotypically resembles more and more the virgin state of the mammary gland. Visualization techniques enable the bioinformatician to extract this type of surprising biological information. Although the resulting image has a convincing biological interpretation, it was not the 'optimal' clustering selected by the particular clustering algorithm. Investigating suboptimal clustering results is essential. First of all, no consensus exists on the meaning of an "optimal" cluster, and secondly, the sheer dimensionality of the space makes it impossible to evaluate the stability of the clusters accurately.

Keeping this in mind, there are several requirements that we would like our clustering algorithm to have:

1. ability to use various dissimilarity measures;

2. ability to deal with many observations;

3. ability to deal with high-dimensional observations;

4. flexibility to investigate suboptimal solutions;

5. ability to visualize resulting clusters.

There are several clustering algorithms that possess this flexibility. Self-organizing maps (SOMs, e.g. Chipman et al. (2003); Tamayo et al. (1999)) is a prototype method that satisfies these requirements, particularly for clustering genes. A SOM imposes a particular two-dimensional geometry of nodes on the data, which then adaptively converge to data clusters. The number of nodes can be altered, and the resulting clustering can be visualized in a two-dimensional profile plot of nodes.

Another worthy method that deserves attention is the mixture model approach proposed by McLachlan et al. (2002). Typically, mixture models are computationally intensive, but by reducing the feature space through elimination of irrelevant genes, the method produces sensible clusterings that allow probabilistic comparisons with alternative clustering proposals.

For clustering samples, hierarchical methods have certain natural advantages. The hierarchical nature of the methods reflects the typical hierarchical nature of the samples. A cancer tissue experiment with replicates has typically several hierarchical layers. First of all, distinct cancers group in distinct clusters. Secondly, within these phenotypical clusters of cancers, several distinct genotypical sub-clusters might be present. Finally, the individuals within these sub-clusters cluster by their replicate tissues.

Several hierarchical methods of clustering exist. The most common one, sometimes simply referred to as *hierarchical clustering*, is a dendrogram method which puts observations together in a pairwise, bottom-up manner. We also consider a novel top-down clustering algorithm, called *hierarchical partitioning around medoids* (HIPAM). A good review of different hierarchical clustering methods can be found in Chipman et al. (2003).

Agglomerative hierarchical clustering

This bottom-up method is a fast, dependable and often-used clustering algorithm that creates the well-known dendrogram (e.g. see Figure 7.2). In a microarray context, it was first used by Eisen et al. (1998) to cluster both the arrays and the genes. This method has been implemented in the Cluster software by the same authors. The original algorithm was described in Cormack (1971).

Although the tree is typically drawn with the branches down, from the point of view of the algorithm it really should have its branches up. It begins by considering all the observations as separate clusters. Then it starts putting together the two samples that are nearest to each other. Any type of distance measure can be used here, although the use of the Euclidean distance is most common. In subsequent stages clusters of observations can also be put together if two of them are close enough. To evaluate the distance between clusters, one can use any of the definitions discussed in Section 7.1.2. From our experience, we believe that average linkage on the log-transformed data leads to the most sensible clusterings. Branches of observations are combined together until eventually one cluster is obtained, that is, the stem of the tree.

Interpretation of a hierarchical tree should proceed with care. The branches of the tree can turn around their supporting axis like a baby's mobile without altering the mathematical structure of the tree. This can make visual inspection of large dendrograms a trying task. Neighbouring nodes are close together only if they lie within the same branch.

To deal with this visual interpretation problem, it has been proposed to slice the tree at a particular level to create distinct clusters. The closer one slices to the stem, the fewer clusters one gets. The height at which to cut the tree can be decided by informal arguments or by some type of cluster validation. Subjective approaches for finding 'interesting clusters' of samples have been applied with some success

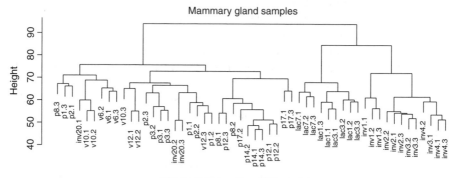

Figure 7.2   The tissues taken from the mammary gland during lactation and early involution separate quite dramatically from the others. It is also clear that late involution is, in terms of gene expression, quite similar to the late virgin state.

(e.g. Hughes et al. (2000)). Although these heuristics may lack formal justification, it is important to remember that clustering is, by its very nature, exploratory.

We apply hierarchical agglomerative clustering to two examples. The **mammary gland experiment** consists of 54 single-channel microarrays. It considers 18 different moments in the development of the mammary gland in mice. Each time point is replicated three times. The 18 time points can be subsumed into four developmental stages of the mammary gland, namely virginity, pregnancy, lactation and involution (period after lactation).

Figure 7.2 shows quite clearly the nested structure of the data: lactation and early involution separate quite clearly from the rest and can be distinguished from each other. Late pregnancy is quite distinct among the others. It is also encouraging to see that replicates mostly cluster closely. This suggests that the experiment was highly replicable, and that the variation as a result of the developmental stage was larger than the technical and biological variation of the replicates.

The breast cancer experiment consisted of 62 microarrays that measured the amplification and deletion patterns of 59 genes in 62 patients. Figure 7.3 shows the hierarchical tree of the 62 patients. Whereas, apparently, a lot of the patients cluster closely together in the center, several distinct clusters are visible.

A possibly problematic aspect of hierarchical clustering is the way it proceeds from the bottom up. Thereby 'mistakes' early on have no way of being corrected later on. Although it does a good job representing the local structure of the data (keeping in mind the invariant permutations of the branches), the global structure might not be represented particularly well. When one has a small number of samples, this is probably not a very serious problem. However, this is an important concern when using hierarchical clustering for finding groups of genes.

Implementation of hierarchical clustering in R. Hierarchical trees are implemented as part of the `cluster` library in R. The function `hclust` creates a hierarchical tree object, whereas the `plclust` function plots it. Hierarchical trees

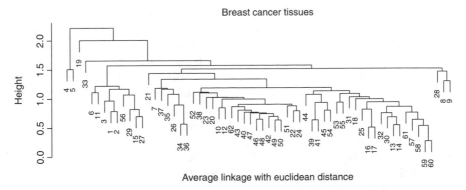

Figure 7.3 Hierarchical clustering of amplification and deletion patterns of 59 genes for 62 breast cancer patients.

can also be plotted as part of the cluster.samples function, implemented in our own library. One needs to specify cluster.samples(data, method= "hierarchical") in order to obtain a dendrogram.

PCA and multiple dimensional scaling

One of the drawbacks of hierarchical clustering is the one-dimensional ordering of the nodes and the imposition of a hierarchical structure even if there may not be any. A way to accommodate these objections—at least partially—is by representing the samples in a two-dimensional plot without imposing a tree structure.

The most common dimension reduction technique is principal component analysis (PCA). PCA is a representation of the original data matrix X into new coordinates. Let us assume that the data matrix X has n rows, representing the samples, and p columns representing the genes. Typically, n is much smaller than p. In order to calculate the principal components, we centre around zero and rescale the columns to have standard deviation equal to one. The 'secret' of PCA is the singular value decomposition (SVD). The SVD of X,

$$X = UDV^t,$$

is a decomposition of the original data matrix into a $n \times n$ matrix U, a $n \times n$ diagonal matrix D with diagonal elements $d_1 \leq d_2 \leq \ldots \leq d_n$, and a $p \times n$ matrix V. Both U and V are unitary matrices. The matrix V is the rotation matrix, which rotates the data X into a new set of coordinates,

$$XV = UD.$$

The matrix UD is a $n \times n$ matrix, where each column is of diminishing importance. Spang et al. (2002) define the columns of UD as *super-genes*. There are only as many super-genes as there are samples. In particular, for prediction and classification super-genes can be useful as they reduce a 'large p, small n' problem to a problem that can be handled with traditional statistical methods.

In order to give a two-dimensional representation of the data, the first two columns of UD are used. The first column is the representation of the data along the first principal component and the second column is the representation along the second principal component. Figure 7.4 shows the principal component projection of the 59-dimensional breast cancer data into a 2-dimensional plane. Clearly this method has advantages over the mere 1-dimensional representation of the branches of the hierarchical tree.

A drawback of the PCA is that the SVD can be rather slow or even computationally impossible, because it operates on the full data matrix. This matrix can have many columns, even though the number of rows may be small. A full PCA for the mammary gland data set requires at least 1 GB of memory and can therefore be perniciously slow.

Multidimensional scaling (MDS) is a collection of methods that does not use the full data matrix but rather the distance matrix between the samples. Typically,

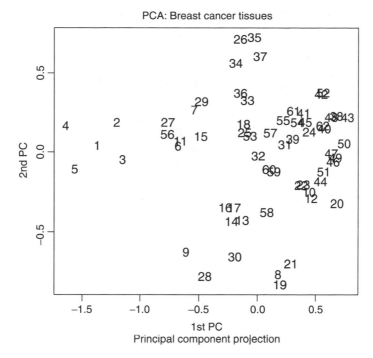

Figure 7.4   Principal component projection of the amplification and deletion patterns for 62 breast cancer patients over 59 genes.

this reduces computation since instead of using the big $n \times p$ matrix, the much
smaller $n \times n$ matrix can be used. One MDS method, Sammon mapping, aims
to find the two-dimensional representation that has the most similar dissimilarity
matrix compared to the original one. The mathematical details of the Sammon
mapping have been discussed in Section 5.2.3.

Figure 7.5(a) shows a representation of the 62 patients that is quite similar to
the principal component representation in Figure 7.4. The advantage of the PCA
representation is that it represents the samples in a Scatter plot whose axes are made
up from (linear combinations of) the most variable genes. The Sammon mapping
treats all genes equivalently. The Sammon mapping is therefore typically 'duller'
than the principal component projection would be.

The Sammon mapping can be applied to data sets of almost any size as long
as the number of samples is limited. It does not have any trouble dealing with the
mammary gland data set, which contains only 54 samples and expression values
for 12,488 spots for each sample.

The use of PCA and SVDs in genomics has been proposed by several people.
Raychaudhuri et al. (2000) showed that the sporulation data set from Chu et al.
(1998) could be summarized to a large extent in only two principal components.
Alter et al. (2000) use SVD to identify eigen-arrays and eigen-genes, which they

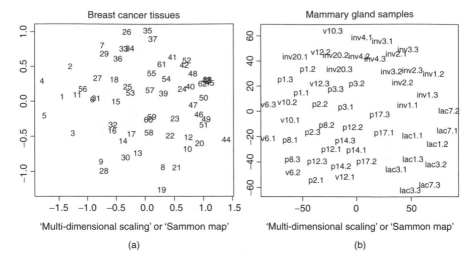

Figure 7.5   Sammon mapping for two data sets: (a) the amplification and deletion patterns for 62 breast cancer patients over 59 genes; (b) the gene expression profiles for 54 mammary gland tissues over 18 distinct time points within four different developmental stages.

subsequently used for analysis. Holter et al. (2000) showed that groups obtained by clustering corresponded in large part with elements in the matrices of the SVD.

Implementation of PCA in R.   Principal component analysis is implemented in R via several functions. The more complete `princomp` has the disadvantage of some numerical instability. We recommend using the function `prcomp` if one is only interested in the principal components, the rotation matrix and the standard deviations. This function is used inside our function `cluster.samples` and can be called via `cluster.samples(data, method="pca")`. Studious people can create their own PCA via the function `svd`, which returns the SVD.

The Sammon mapping is included in the library `MASS` as the function `sammon`. We made use of this function inside our own routine `cluster.samples`. To create a Sammon map of the data, use `cluster.samples(data, method="sammon")`.

Hierarchical PAM

Both hierarchical clustering and forms of MDS have important benefits for clustering samples, involving microarray data. Hierarchical clustering creates an easily interpretable tree-type structure, whereas MDS gives a better insight into the geometrical structure of the data. Both of them have some disadvantages as well. There are some problems associated with the bottom-up nature of hierarchical clustering, whereas multidimensional scaling does not actually come up with a proposed class label.

In this section, we discuss a novel method that tries to combine the strengths of the hierarchical and purely visual methods. It is based on the PAM algorithm by Kaufman and Rousseeuw (1990) and has close similarities with the DIANA (divisive analysis) algorithm from the same authors and the HOPACH (hierarchical ordered partitioning and collapsing hybrid) algorithm recently proposed by Pollard and van der Laan (2002).

PAM is a clustering algorithm akin to k-means clustering. PAM, as well as k-means, require the user to specify the number of clusters k in advance. Just like k-means, the objective of the clustering is to select a partition $\{C_1, C_2, \ldots, C_k\}$ of sample clusters *and* a set $\{m_1, m_2, \ldots, m_k\}$ of cluster centres to minimize,

$$\sum_{i=i}^{k} \sum_{j \in C_i} d(x_j, m_i),$$

where d is some distance measure of interest. k-means typically uses the Manhattan distance or the squared Euclidean distance, although this is not essential. Any symmetric, non-negative measure can be used. The difference between k-means and PAM is that for k-means any choice of cluster centres $m_1, \ldots, m_k$ is allowed, whereas in PAM this choice is restricted to the set of observations $x_1, \ldots, x_n$. This restriction has important advantages. The PAM algorithm is much more stable, whereas the k-means algorithm sometimes converges to some local optimum. Moreover, PAM works much faster in high-dimensional problems, for instance when clustering microarray data.

There are two possible disadvantages of clustering algorithms, such as PAM and k-means. First of all, the clustering is non-hierarchical. Especially, in cases where such hierarchical structure is believed to exist, this might be considered a drawback. Furthermore, these methods have no way of evaluating the total number of clusters. For each user-specified number of clusters k, it can find the best possible clustering, but it cannot compare different choices of k.

Hierarchical partitioning around medoids (HIPAM) is designed to deal with these issues. It is a hierarchical top-down method, designed in order to overcome the disadvantages of a bottom-up method. The global clustering structure of a top-down method is more stable than one can expect from a bottom-up method. Just like DIANA, HIPAM uses PAM to partition the data at each subsequent level. However, whereas DIANA does not have any stopping rule and divides the data until the lowest level of the individual data points, HIPAM has a way to evaluate the validity of a particular clustering. It uses an accepted scoring rule to evaluate the clustering in order to decide whether or not to continue down the tree. This scoring rule is based on the average silhouette width (a.s.w.), proposed by Kaufman and Rousseeuw (1990). HIPAM can maximize a.s.w. globally, or it can evaluate the improvements of the a.s.w. in each of the branches locally. This latter method has been proposed by Pollard and van der Laan (2002).

Figure 7.6 shows the results of a HIPAM clustering of the breast cancer data set. The 62 patients, for whom the amplification and deletion patterns of 59 genes

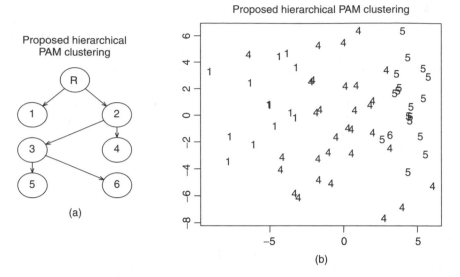

(a)

(b)

Figure 7.6   Hierarchical PAM clustering of the breast cancer tumour data. Ultimately, four clusters, 1, 4, 5 and 6, are selected. The global a.s.w. is 0.135.

were measured, are clustered in four separate clusters. Clusters 5 and 6 are part of cluster 3, which together with cluster 4 is nested in cluster 2. Figure 7.6(b) shows how the class labels are distributed over the patients, via a two-dimensional Sammon plot.

Cluster validation: global.   There are several ways to measure the validity of a clustering. The most obvious statistical choice would be a measure based on the likelihood. However, such a measure would presuppose some kind of statistical model and can be rather computationally intensive for large data sets. Instead, we consider a non-parametric measure, which allows us to compare different clusterings easily. It is called the *average silhouette width* (a.s.w.) and was proposed by Kaufman and Rousseeuw (1990).

Given a particular clustering $\{C_1, \dots, C_k\}$, one defines for each observation $x_i \in C(x_i)$ a 'distinguishability' score, called the *silhouette width*:

$$s(x_i) = \frac{b_i - a_i}{\max\{a_i, b_i\}}, \tag{7.2}$$

where

$$a_i = \frac{1}{|C(x_i)|} \sum_{x_j \in C(x_i)} d(x_j, x_i),$$

$$b_i = \min_{C \neq C(x_i)} \frac{1}{|C|} \sum_{x_j \in C} d(x_j, x_i).$$

It is easy to show that s_i lies between -1 and 1. The higher the value of s_i, the better the clustering of x_i in $C(x_i)$ is. By taking the average over all s_i, one gets a global validity measure of the clustering. This measure is the average silhouette width,

$$\text{a.s.w} = \bar{s}.$$

One form of HIPAM can continue splitting clusters, until a maximum a.s.w. is reached.

Figure 7.7(a) shows an application of global a.s.w. optimization using hierarchical PAM in the example of the 54 mammary gland tissues in four developmental stages. It reaches an a.s.w. maximum of 0.244 with only two classes, separating tissues in the lactation and early involution phase from the rest.

It is not uncommon that the a.s.w. finds only a small number of classes when a lot of data is present. The reason is that more data will obscure the boundaries between classes and lead to very low silhouette values on these boundaries. Although the separation between the lactation and early involution phase from the rest is interesting by itself, it might be helpful to be able to split the dendrogram further. The simplest way is by investigating suboptimal solutions. Figure 7.7(b) shows the largest suboptimal solution less than 0.1 away from the tree with the optimal a.s.w.

A disadvantage of the global HIPAM method is that the dendrogram found might depend on the order in which the branches are grown. Since the method is a greedy search method, growing the tree in one branch first might prevent it from growing in another branch later.

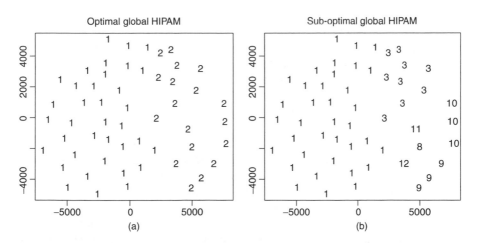

Figure 7.7 Hierarchical PAM clustering of the Mammary Gland tissues; (a) the optimal solution (a.s.w. = 0.244) gives two clusters, (b) whereas a suboptimal alternative (a.s.w. = 0.148) yields seven.

Cluster validation: local. The local HIPAM method is proposed to overcome two weaknesses of the global HIPAM method. First of all, optimizing the global average silhouette width (a.s.w.) tends to result in only a few clusters in large data sets. Maximizing the global a.s.w. is a bit like looking at the earth from space. Typically one can only distinguish sea from land. However, when one gets a bit closer and focuses on land only, then it becomes possible to distinguish cities, farmland and hills. This is the idea behind the local HIPAM method. It continues to grow a branch until it reaches some local homogeneity. In this way, local HIPAM does not suffer from the second weakness of the global method: the method is independent from the way the branches are grown.

For this algorithm, one uses the a.s.w. only locally as a measure of homogeneity. The a.s.w. of the best possible split can be interpreted as a level of homogeneity of the mother node. The higher the a.s.w., the lower the homogeneity of the mother node. By comparing this homogeneity with the average homogeneity of the child nodes, one has a decision rule whether or not to split. If the homogeneity of the mother node is less than the average homogeneity of the children, then the split is accepted. In other words, one accepts a split if

$$\text{a.s.w.(mother)} > \frac{1}{k} \sum_{i=1}^{k} \text{a.s.w.(child}(i)).$$

Variations on this theme are possible. One can, for instance, introduce a penalty term, which only allows a split if the a.s.w. of the mother node is a certain amount more than the *mean average silhouette width* of the children. The idea of evaluating the branches via the mean average silhouette width comes from the HOPACH algorithm (Pollard and van der Laan 2002). It is also possible to accept a split only if the a.s.w. of the mother node is larger than a pre-specified constant.

In Figure 7.8, the local validation algorithm is applied to the breast cancer DNA amplification and deletion data and the mammary gland expression data. The two-dimensional layout of the clustering shows interesting details. The mammary gland tissues, for example, clearly separate into the involution phase (top), the lactation phase (bottom left) and the rest (right).

Implementation of HIPAM in R. Hierarchical PAM is implemented in R as the function `hipam`. It requires a data matrix for which the rows are to be clustered. By specifying either the `method = "local"` or the `method = "global"`, HIPAM will either use the global a.s.w. optimization criterion or the local branch splitting method. We recommend using the `"local"` method, which is set as the default. By setting the `asw.tol` constant different from zero, one can introduce a tolerance (positive `asw.tol`) or penalty (negative `asw.tol`) in the branch splitting procedure. The value of `asw.tol` is ignored, however, if the local method is selected *and* a value for `local.const` is specified. In that case, branches will be split until their associated local average silhoutte widths get below the specified value of `local.const`. Only values between -1 and 1 are meaningful choices for `local.const`.

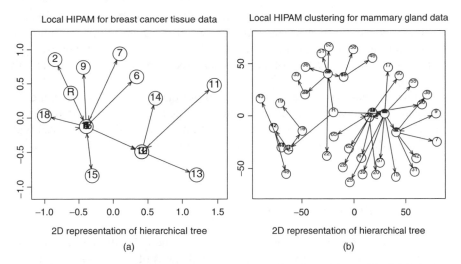

Figure 7.8   Two-dimensional dendrograms resulting from hierarchical PAM clustering using local cluster validation. The root node is indicated by R. Some cluster centres overlap; (a) The 62 breast cancer tumours split into 11 different clusters, (b) whereas the 54 mammary gland tissues eventually divide into 38 clusters. The clusters on the right correspond to the virgin, pregnancy and late involution phase. The top clusters correspond to the early involution phase, whereas the clusters on the left contain the tissues in the lactation phase.

The result of the function `hipam` is a HIPAM object. This object can be visualized with the help of the `plot.tree` command. The `plot.tree` command can plot a hierarchical tree, `method = "tree"` (with or without `reordered` branches), a two-dimensional representation of this tree, `method = "2d"`, or a 2D-representation of the cluster labels of the observations, `method = "all"`. For the latter representation of the clustering, it needs to be supplied with the original `data` as well.

7.2   Exploratory Supervised Learning

The explosion of clustering algorithms has sometimes obscured the fact that clustering by itself is not necessarily meaningful. Typically, only when the scientist is able to link the observed clusters of microarray samples with some phenotypical properties of the underlying tissues, the clustering is considered successful. As a simple example, we could say that a microarray experiment is replicated very well if replicate samples cluster together. More interesting would be the case in which we can link, for example, survival information with the clustering of microarray data of the tumours.

Linking covariate data with some response of interest is the field of supervised learning. However, many supervised learning techniques are formal and do not admit easy visualization. Exploratory supervised learning is not a field that is particularly developed. Nevertheless, it is easy to extend clustering algorithms to include relevant visual information relating to the variable of interest. For example, it is possible to label the leaves of a hierarchical tree with the value of any variable of interest.

Ultimately, supervised methods, such as those discussed in Chapter 9, are the best means to evaluate the relationship between gene expression patterns and a response of interest. Visual inspection does not preclude or exhaustively deal with such relationships, although it can reveal new insights in the data for the empirical scientist.

7.2.1   Labelled dendrograms

Figure 7.9 shows the same hierarchical tree as in Figure 7.3. In the new dendrogram, the leaf labels are replaced by the survival status of the individual patients. If clusters of labels appear close together, then it might suggest that the overall expression pattern is related to the class labels. This might suggest that using the data we might have a good chance to classify the classes based on the expression data alone.

The reality is frequently more complicated. As can be seen in Figure 7.3, there is no overwhelming evidence that death due to breast cancer separates clearly according to the expression pattern of all the 59 genes. This is perhaps not too surprising. It could be that some of the genes might not be related to the seriousness of the cancer, or it could be that there are simply several expression subtypes, each of which relates to a particular survival profile.

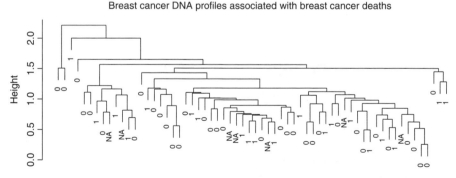

Breast cancer DNA profiles associated with breast cancer deaths

Of the 62 patients in the study, 20 died of breast cancer (1), whereas 35 didn't (0); for 7 no data were available (NA)

Figure 7.9   The leaves of the dendrogram are labelled with the survival status of the patients; (0: alive, 1: dead).

7.2.2   Labelled PAM-type clusterings

In principle, it is very easy to study the distribution of a particular covariate within each cluster. Each of the clusters represents, for example, a group of patients. For each group, the distribution of a particular variable of interest, for example breast cancer deaths, can be calculated. By comparing the distributions across the clusters, one can judge informally whether differences in gene expressions might be responsible for differences in the breast cancer deaths.

The result of any 'PAM-type (or k-means) clustering is a set of cluster labels for all the samples and a set of medoids (or centroids) for each of the clusters. In this section, we propose to visualize the clusterings by means of multidimensional scaling of the cluster medoids. We then superimpose the distribution of the response variable in each of the clusters onto the 2D-scaled medoids.

Binary response variable

Binary response variables are common in many applications. In biological experiments, it can stand for wild-type versus mutant, transgenic versus control or something else. In medical applications, it could stand for survival versus non survival. This is the case in the breast cancer experiment, in which we have information about breast cancer survival for 55 patients out of 62.

Figure 7.10 shows the 5 patient clusters with their associated breast cancer survival status. There is no obvious difference between the different clusters. Only cluster 4, perhaps, has a slightly worse breast cancer survival record.

Multinomial response variable

A multinomial response variable is an outcome variable that can take more than two possible values. There are two main types of multinomial variables: ordinal and nominal. Unordered or nominal multinomial variables are typically class or group labels, such as disease type or treatment, whereas ordinal multinomial variables are ordered sets of levels, such as developmental stage or histological grade.

In the breast cancer data set, we have available the histological grade of each tumour. Histology is the study of tissues, including cellular structure and function. Pathologists often assign a histologic grade to a patient's cancerous breast tumour to help determine the patient's prognosis. The Scarff-Bloom–Richardson system is the most common type of cancer grade system used today. To determine a tumour's histologic grade, pathologists closely observe three features of the tumour: the frequency of cell mitosis (rate of cell division), tubule formation (percentage of cancer composed of tubular structures) and nuclear pleomorphism (change in cell size and uniformity). Each of these features is assigned a score ranging from 1 to 3 (1 indicating slower cell growth and 3 indicating faster cell growth). The scores of each of the cells' features are then added together for a final sum that will range from 3 to 9. A tumour with a final sum of 3, 4, or 5 is considered a Grade 1 tumour (well differentiated). A sum of 6 or 7 is considered a Grade

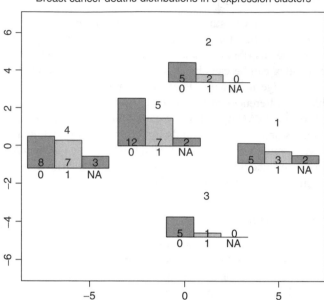

Figure 7.10   Death due to breast cancer for 5 expression clusters of 62 patients; (NA: no data available, 1: breast cancer death, 0: otherwise).

2 tumour (moderately differentiated), and a sum of 8 or 9 is a Grade 3 tumour (poorly differentiated). Lower grades are shown to correlate positively with breast cancer survival.

Continuous response variable

In the breast cancer example, one can think, for instance, about the Nottingham prognostic index (NPI), which is based on tumour size in breast, node involvement and Scarff–Bloom–Richardson grading (Galea et al. 1992). It has been shown to be of prognostic value for the treatment of the breast cancer. In the original paper, the following prognosis was suggested: good: NPI $\leq$ 3.4; moderate: 3.4 < NPI $\leq$ 5.4; and poor: NPI > 5.4. Figure 7.12(a) does not suggest any differences in NPI between the clusters (Chollet et al. 2003).

Figure 7.12(b) shows that the tumours in class four are larger on average than any of the other classes. This, rather than any of the expression differences between the clusters, might be the real explanation why the number of breast cancer deaths in class four is higher than in any of the other clusters.

Implementation of labelled PAM-type clustering in R.   Plots such as in Figures 7.11 and 7.12 can be obtained in R using the `sammon.plot` function. The user needs to supply a clustering object from a PAM-type clustering (e.g. PAM, HIPAM

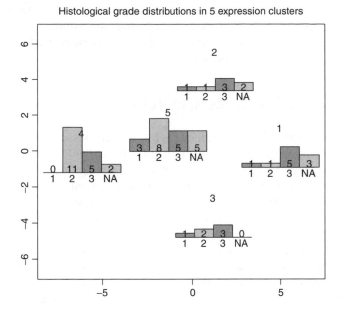

Figure 7.11   Histological grade distribution for 5 expression clusters of 62 patients.

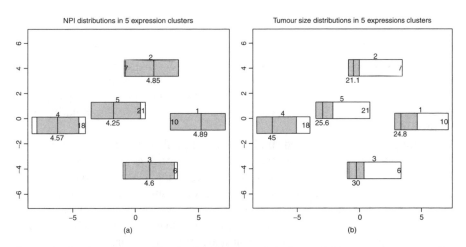

Figure 7.12   (a) Average Nottingham prognosis index and (b) average tumour size for 5 expression clusters of 62 patients; the shading indicates one standard deviation.

or PAMSAM) and a covariate of choice. The length of the covariate vector needs to be as long as the number of samples. The function `sammon.plot` interprets both numerical and categorical data. If angular data is supplied, then the option `method="circular"` should be selected.

7.3   Discovering Gene Clusters

As soon as microarrays became available in the late nineties, one of the first desires was to reduce the large quantity of information for individual genes into groups of similarly expressed genes. This was partially an attempt to make the data more manageable and partially to see if genes of similar known function have similar expression profiles. Early successes in relating gene expression to cellular function (Eisen et al. 1998) suggested a simple means of gaining insight into the function of those genes for which information is not available.

There have been a large number of contributions to the field of clustering gene expressions. Roughly, we can divide the field into four groups: (i) hierarchical clustering, (ii) centroid-type clustering, (iii) clustering based on the SVD and (iv) model-based clustering. Several comparative studies (Chipman et al. 2003; Datta and Datta 2003) discuss the advantages and disadvantages of some of these methods. It is our aim in this section to describe several common and useful clustering algorithms and apply them to several data sets.

7.3.1   Similarity measures for expression profiles

Typically, several gene expression values are observed for each gene in a microarray experiment. The vector of gene expressions for a gene is typically referred to as the *expression profile* of that gene. Expression profiles are typically stored as columns in a gene expression matrix. In order to cluster genes together, it is necessary to determine for each pair of genes the similarity or dissimilarity of the expression profiles.

Although we have discussed similarity and dissimilarity measures before (Section 5.2.2, Section 7.1.2), the meaning of 'vicinity' of gene expression profiles is different from that of other types of objects. For example, whereas two gene expressions that vary in a coordinated fashion but in opposing directions could suggest that the underlying genes are related, such an observation would not make sense if we compared two microarray samples; two microarrays for which expression values are opposed suggest very different underlying conditions.

Co-expression and co-regulation

The term *co-regulation* (Gasch and Eisen 2002) refers to common regulatory control of genes in order to fine-tune an organism's response to varying conditions. It is thought that co-regulation simplifies the collaboration of gene products namely proteins. Co-regulation of genes can be measured via coordinated expression, or *co-expression*, of these genes in different circumstances.

There are several mathematical definitions of co-expression. The most obvious one is to call two genes co-expressed if they have a high *absolute correlation*. The dissimilarity between two gene expression profiles x and y can then be defined as,

$$d_1(x, y) = 1 - |r(x, y)|,$$

where $r(x, y)$ is the Pearson sample correlation coefficient. Both strongly positively correlated as well as negatively correlated genes are considered co-expressed.

The sample correlation coefficient can be calculated on (i) the original scale, (ii) the log scale or (iii) the rank scale. In the latter case, it is also known as *Spearman's rank correlation*. As the correlation coefficient measures the amount of *linear* correlation between the data *and* expression data can be rather noisy with large outliers, we suggest to avoid using the data on the original scale for clustering purposes. In principle, the log scale stabilizes the variances and can be used by default. If strong suspicions about the validity of the data remain, ranked data can be used to make conclusions more robust.

Eisen et al. (1998) use a variation of Pearson correlation to define co-expression. They consider the cosine measure,

$$d_2(x, y) = 1 - S(x, y)$$

$$= 1 - \frac{\sum x_i y_i}{\sqrt{\sum x_i^2 \sum y_i^2}}.$$

The cosine measure represents the angle between the expression profiles x and y relative to the absolute zero point. This measure is therefore only appropriate if the values x_i and y_i represent log-expression ratios relative to some control. This is not always the case. Another aspect of d_2 is that it considers negatively related expression profiles as extremely different. It is therefore safer to use a correlation measure.

Other dissimilarity measures

Typical geometric distance measures, such as Euclidean and Manhattan distance, can also be used. The advantage is that they are readily implemented, but their interpretation is not so much 'co-expression' as merely 'similar profiles'. Whereas the absolute correlation similarity measure judges all four profiles in Figure 7.13 as similar, the geometric measures only considers profiles 1 and 2 as similar and would not cluster them together with profiles 3 and 4.

Although we prefer using the correlation or absolute correlation for measuring gene similarity, there are sensible ways to apply geometric measures to model driven parameters fitted to the expression data.

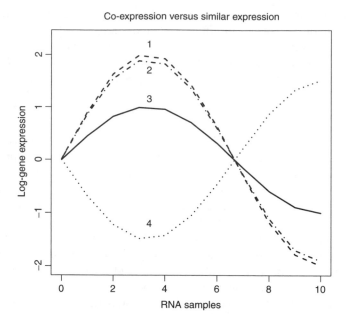

Figure 7.13   Whereas all four expression profiles come from co-expressed genes, only two of them have a similar profiles.

If the microarray experiment is a time-course experiment, then the expression data are interpreted as time-series data. Let

$$x = (x_{t_1}, x_{t_2}, \ldots, x_{t_n})$$

represent the gene expression profile of a particular gene measured at times $t_1, \ldots, t_n$. A way to describe the temporal structure of the data is by modelling the data in a time-series model. A popular model is the Box–Jenkins autoregressive-moving-average (ARMA) model (Box et al. 1994). It allows the dynamic structure and the error process of the expression values at a particular time to depend on previous time points. More precisely, the ARMA(p, q) model is defined as

$$x_{t_i} = \sum_{j=1}^{p} \alpha_j x_{t_{i-j}} + \sum_{k=0}^{q} \beta_k \epsilon_{t_{i-k}}. \tag{7.3}$$

By replacing each gene expression vector x in the clustering algorithm with the estimated parameter vector,

$$\hat{\theta} = (\hat{\alpha}_1, \ldots, \hat{\alpha}_p, \hat{\beta}_1, \ldots, \hat{\beta}_q),$$

one can continue to use the geometric distances to evaluate the similarity of the underlying two expression profiles. Take for example the data in Figure 7.13. We

Table 7.1 Parameter estimates of an ARMA(0,3) model applied to data in Figure 7.13.

	$\hat{\beta}_1$	$\hat{\beta}_2$	$\hat{\beta}_3$
1	2.59	2.59	1.00
2	2.47	2.47	1.00
3	2.57	2.57	1.00
4	2.60	2.60	1.00

fit an ARMA(0,3) process to each of the expression profiles, and the estimated moving average parameters are displayed in Table 7.1. Although the expression profiles are very different, the parameter estimates are very similar. By defining the dissimilarity between profiles as the Euclidean distance between the parameter estimates, one can obtain more meaningful results.

This approach shares similarities with the CAGED algorithm proposed by Ramoni et al. (2002). They suggest modelling each dynamic gene expression profile by some autoregressive (AR) process. Their comprehensive Bayesian analysis selects the appropriate order of the process in a data-driven fashion and is less *ad hoc* than what we have suggested. Dougherty et al. (2002) take a similar view. They suggest that meaningful clusters should represent a partition of the gene profiles according to the process to which they belong. They implement their ideas via a model-based clustering algorithm.

7.3.2 Gene clustering methods

Gene expression profiles have often been clustered using hierarchical clustering. Hierarchical clustering produces a tree, a so-called *dendrogram*. Dendrograms are excellent for expressing hierarchical structures such as evolutionary relationships or subtypes of larger classes. Although they have clearly proven valuable to deduce interesting biological conclusions, there are several drawbacks to the method. Dendrograms lack biological interpretation when clustering expression profiles. Moreover, agglomerative hierarchical clustering makes decisions based on local features from the bottom-up without the possibility of any revision later. This means that global clustering structures high up in the tree are notoriously unreliable, whereas these global features are of particular interest to the biologist. We have previously mentioned the non-unique representation of a dendrogram. Because of these reasons, we would not generally recommend agglomerative hierarchical methods for clustering genes.

Hierarchical clustering: agglomerative and two-way PAM

Ordinary hierarchical clustering has been the gene profile clustering algorithm of choice in applications (Lee et al. 2003). It represents the genes in a hierarchical

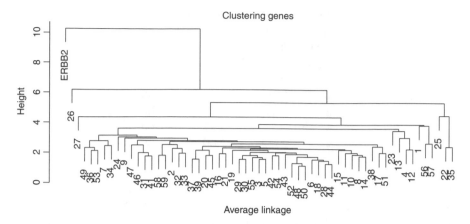

Figure 7.14   A dendrogram of the 59 amplification and deletion profiles of breast cancer data set.

fashion, like a tree. It puts genes with similar expression patterns in the same branches. The results, a dendrogram, can give the scientist an immediate impression of groups of genes as well as of genes that are very similar. Figure 7.14 shows the hierarchical tree of the breast cancer data set using an absolute correlation distance measure, where the leaves represent the genes. One thing immediately clear from this plot is that the growth factor receptor ERBB2 has a gene amplification pattern that is quite different from the other genes. It is known that amplification of ERBB2, which is a growth promoter, leads to increased cell division and may be involved in the formation of cancerous cells (Slamon et al. 1989).

To produce Figure 7.14, the absolute correlation similarity was used. This is not a typical default for agglomerative hierarchical clustering algorithms. However, it expresses more clearly what it means for two genes to be closely related. The other technical details of the hierarchical clustering algorithm have been discussed before in the context of clustering samples (*cf.* Section 7.1.3).

Hierarchical clustering has been applied in quite diverse areas of microarray data analysis ranging from searching for generic networks in the yeast cell cycle (DeRisi et al. 1997) to the prediction of clinical outcome of breast cancer (Van 't Veer et al. 2002). One of the appeals of hierarchical clustering, as illustrated for example in Eisen et al. (1998), is the way it reorganizes the expression matrix into an organized set of patterns. A rectangular matrix with colours ranging from green via black to red represents the microarray data matrix of a complete experiment. By looking for peculiar patterns and patches in this matrix, the scientist hopes to find ways to relate groups of gene expression profiles to, for example, clinical outcome.

However, there are other ways to create similar plots that can be even more effective. The ordering of the leaves in a hierarchical clustering is not necessarily the best way to order the many gene columns of the data matrix. It is hard to spot the boundaries between 'clusters', and the order itself is not optimal for large numbers of objects.

Other approaches to two-way clustering have been proposed. Getz et al. (2000) propose a coupled two-way clustering algorithm. The idea is to find subsets of genes and samples such that if one is used to cluster the other then stable and informative partitions emerge. Plaid models have been suggested in Lazzeroni and Owen (2002). These allow a gene to be in more than one cluster or in none. Also, clusters of genes can be defined with respect to only a subset of samples. Several genes might cluster together in a particular set of conditions, whereas under other conditions they do not cluster together. These are some useful developments in the field of two-way clustering.

We propose a **two-way clustering algorithm based on PAM**. First, the samples are clustered via a PAM-based algorithm using the full gene expression matrix. Then the genes are clustered according to their profiles with the possibility that they are omitted if they do not distinguish between the sample clusters. This latter feature makes this algorithm similar to coupled clustering.

Figure 7.15 shows the two-way PAM clustering for the breast cancer data. The 62 samples are clustered into five groups, whereas the 59 genes are clustered into six groups. The sixth group of 10 genes is not shown as it was judged uninformative for distinguishing between the sample clusters. The tumour clusters can then be related to clinical outcome, as in Table 7.2. One could speculate, for example, whether the difference in breast cancer deaths between tumour classes 2 and 3 might be related to the marked difference in amplification patterns in gene clusters 4 and 2.

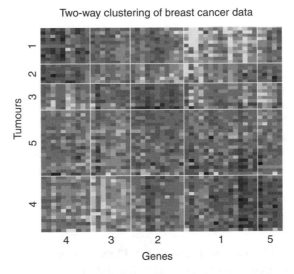

Figure 7.15   Two-way clustering of gene amplification data in the breast cancer experiment.

Table 7.2   Clinical outcomes, and their standard deviations in brackets, associated with the two-way PAM clustering in Figure 7.15.

Variable	Sample class				
	1	2	3	4	5
Number of tumours	11	6	8	17	20
Breast cancer deaths	3/9	1/6	4/8	6/14	6/18
Survival time (years)					
Uncensored	4.7 (3)	0.1 (–)	2.1 (2)	1.9 (2)	3.6 (3)
Censored	5.9 (3)	6.5 (2)	8.1 (1)	6.3 (5)	7.2 (3)
Tumour size (mm)	24 (18)	30 (8)	21 (8)	46 (29)	27 (11)
NPI	4.8 (2)	4.6 (1)	4.5 (1)	4.6 (1)	4.4 (1)
Grade	2.6 (1)	2.3 (1)	2.0 (1)	2.3 (1)	2.2 (1)

The algorithm for this type of two-way clustering is based on the PAM algorithm (Kaufman and Rousseeuw 1990). It finds a number of tumour clusters, $C_1^s, C_2^s, \ldots, C_k^s$, that maximizes the average silhouette width (a.s.w.) using PAM at each step. Genes are considered informative for this tumour clustering if their between-sums-of-squares significantly outweighs the within-sums-of-squares. Put differently, the following model is considered for the log-expression values y_{ij} for each gene i and condition or tumour j,

$$y_{ij} = \alpha_{C^s(j)} + \epsilon_{ij}, \quad j = 1, 2, \ldots, n,$$

where $C^s(j)$ is the class in which tumour j has been clustered. From standard theory

$$\frac{\frac{1}{k-1} \sum_{j=1}^{n} (\hat{\alpha}_{C^s(j)} - \overline{y}_{i.})^2}{\frac{1}{n-k} \sum_{j=1}^{n} (y_{ij} - \hat{\alpha}_{C^s(j)})^2} \sim F_{k-1, n-k},$$

approximately, under the null hypothesis that gene i is uninformative. For each gene, a test can be performed at the required level of significance. It should be noted that many genes will appear significant simply because the tumour clusters were selected on the basis of these particular genes. Therefore, the usual interpretation of the p-value is not applicable here. The test should mainly be regarded as rooting out those genes that certainly are not informative. For the remaining genes, the clusters are determined via another PAM clustering coupled with maximizing the average silhouette width.

Implementation of two-way PAM in R.   Plots such as in Figure 7.15 can be obtained both in colour and in grey-scale in R using the twoway.pam function. The user needs to supply the gene expression matrix, with the genes as the columns and the samples as the rows. One can specify the number of row and column clusters via the `k.row` and `k.col` parameters. If they are left unspecified, then `twoway.pam` searches for the 'best' values. The option `reduce.cols` can be selected in order to eliminate genes that are uninformative in the resulting sample clustering.

PAMSAM: a partitioning method

In this section, we discuss a new kind of partitioning method, called PAMSAM. It combines a k-means-type clustering method with a method for visualizing the levels of similarity between clusters via a Sammon plot. Unlike most other clustering methods, including PAM and k-means, it does not fix the number of clusters in advance. It selects them via the average silhouette width criterion. After selection of the clusters, PAMSAM visualizes the cluster centres relative to one another, as in a SOM. Similar clusters can be put together manually, in case this is required.

Partitioning methods differ from hierarchical methods in that they try to partition the set of objects in such a way that a particular score function has been optimized. k-means and PAM (k-medoids) aim to find clusters $C_1, \ldots, C_k$ with associated centroids or 'cluster centres' $m_1, \ldots, m_k$, such that the sum of *within-cluster sums of squares* is minimized,

$$(C, m) = \arg \min \sum_{i=1}^{k} \sum_{x_{ij} \in C_i} (x_{ij} - m_i)^2. \tag{7.4}$$

k-means and related methods are popular ways to cluster genes into (a pre-specified number) k classes. Chipman et al. (2003) notice that this is a *combinatorial optimization problem*. Given that the number of genes is typically large ($\sim$10,000) in microarray settings, it is impossible to check each partitioning of k classes.

Worse still, k-means and PAM do not have any way of checking the appropriate value for k. Minimizing Equation (7.4) over k leads to an unsatisfactory choice of k equal to the number of observations. Kaufman and Rousseeuw (1990) suggested another score that naturally penalizes for too large choices of k—the average silhouette width (a.s.w.). The a.s.w. for a particular clustering $C = (C_1, \ldots, C_k)$ is defined as

$$\text{a.s.w.}(C) = \frac{1}{p} \sum_{i=1}^{p} \frac{b_i - a_i}{\max\{a_i, b_i\}}, \tag{7.5}$$

where

$$a_i = \frac{1}{|C(x_i)|} \sum_{x_j \in C(x_i)} d(x_j, x_i),$$

$$b_i = \min_{C \neq C(x_i)} \frac{1}{|C|} \sum_{x_j \in C} d(x_j, x_i),$$

and x_i is the gene profile for gene i. This score has been used before in Section 7.1.3. The average silhouette width has been shown to have good performance in distinguishing clusters (Bryan et al. 2002). Large values of a.s.w. correspond to clear differences between clusters.

However, maximizing a.s.w. explicitly is an even more difficult problem than minimizing the within-cluster sums of squares. PAMSAM is an algorithm that searches an approximate answer to the maximal a.s.w. by using the speed of the PAM algorithm.

The PAM-clustering algorithm detects a pre-specified number of clusters, k. It creates clusters by selecting k gene profiles $m_1, \ldots, m_k$ from the original data set (rather than from a general high-dimensional space as k-means does), called *medoids*, and assigning the observations to the closest medoid such that

$$\sum_{i=1}^{k} \sum_{j \in C_{m_i}} d(m_i, x_j)$$

is as small as possible. By restricting the search of the cluster centres to the observed profiles, PAM is more robust to outliers than other centroid-based methods, such as k-means. Moreover, this restriction allows the method to be much quicker than techniques that have to search through some high-dimensional space. If the number of objects is very large, a sampling-based variation of PAM known as CLARA (clustering large applications) (Kaufman and Rousseeuw 1990) can be used to speed up the search.

For each of the PAM-clustering proposals, an a.s.w. score can be calculated. By reassigning objects from one class to another locally, PAMSAM attempts to increase the a.s.w. This is done via a number of expand-and-collapse steps within the algorithm. The final PAMSAM clustering is the one that attains the highest a.s.w.

The relative similarity of the selected PAMSAM clusters is visualized in two dimensions using a particular form of multidimensional scaling, known as Sammon mapping (Sammon 1969). Sammon mapping creates a representation of objects in lower-dimensional space that attempts to minimize the extent to which the distances between each of the objects is distorted. The cluster medoids are used to represent each cluster in this 2D-plot.

Figure 7.16 shows the PAMSAM plot associated with mammary gland experiment. In this experiment, 12,488 genes and ESTs were measured across 18 different stages of the mammary gland development. The 5,657 genes in clusters 3,5 and 7 represent genes that show very little change across the whole time course. Cluster 4 contains 1,574 genes that are up-regulated during pregnancy. The 1,636 genes in cluster 1 are up-regulated during lactation. It is known that lactation is a process that has enormous repercussions on the system. The mother switches off many other bodily processes during lactation. It is therefore not surprising to see that there is a large number of genes (3,079 in cluster 2 and 1,574 in cluster 4) that show significant down-regulation during this period. Finally, there is a class of 1,542 genes in cluster 6 that show marked up-regulation during the final involution stage.

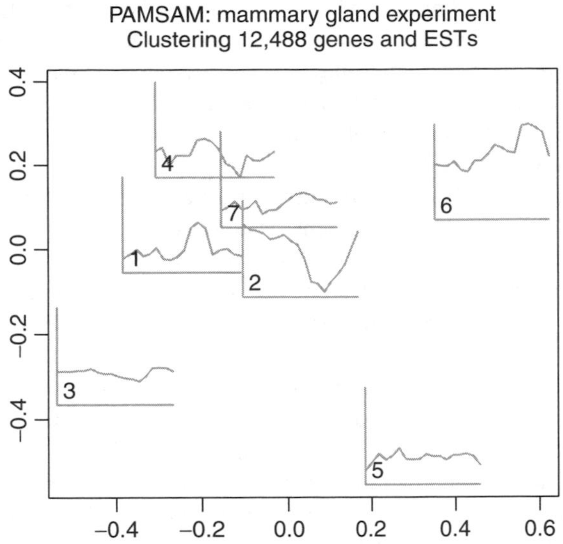

Figure 7.16   The medoids of the seven gene clusters, selected via PAMSAM.

Implementation of PAMSAM in R.   The PAMSAM algorithm has been implemented in R as the pamsam function. The function pamsam requires a data matrix in such a form that the rows are the objects to be clustered. If a value for k is specified, it will perform a PAM or CLARA clustering for k clusters. The option metric allows the user to specify a number of different dissimilarity measures used for the clustering. Currently implemented are the Euclidean ("euclidean"), Manhattan ("manhattan"), L_3 ("cubic"), L_4 ("quartic"), correlation ("correlation") and absolute correlation ("abscor") dissimilarity measures.

SVD clustering: super-genes and gene shaving

The SVD is a powerful tool in high-dimensional data analysis to find principal directions in large data sets. Typically, there are many more genes than there are samples in a microarray experiment. The full data matrix X has many more columns, p, than rows, n. The SVD of X,

$$X = UDV^t,$$

is a decomposition of the full $n \times p$ data matrix into a $n \times n$ matrix U, a $n \times n$ diagonal matrix D with diagonal elements $d_1 \le d_2 \le \ldots \le d_n$, and a $p \times n$ matrix V. Both U and V have orthonormal rows and columns. The matrix V is the rotation matrix, which rotates the data X into a new set of coordinates,

$$XV = UD.$$

The matrix UD is a $n \times n$ matrix, where each column is of diminishing importance. Rather than using the original columns of X, Mike West and co-workers suggest to use the columns of XV, which they term *super-genes* (Spang et al. 2002). The transformed data matrix XV is much smaller than the original matrix with equal number of observations and variables and is therefore much more amenable to traditional statistical methods.

Gene shaving is another SVD-based technique, which has been put forward in Hastie et al. (2000) as an improvement on traditional clustering techniques. The aim is to find clusters with small numbers of similar genes that 'vary as much as possible across the samples'. Given that most genes are unlikely to be doing much in any conditions being considered, this would seem a sensible aim in many situations.

Clusters are found sequentially via the repeated application of the SVD, also called principal component analysis in statistics. To find the first cluster, the main principal component of the data is found—this is the first column of the matrix XV, that is, the linear combination of genes with the largest possible variance. The inner product of each gene and the first super-gene is calculated. Some fraction of the least correlated genes are removed from the data X to create a new data matrix X'. This is repeated until some criterion, called the gap-statistic, has been optimized.

Once the first gene shaving cluster is chosen, the data are then orthogonalized to the 'average' gene from the first cluster. The process is repeated again on this orthogonalized data: the first principal component is found and the genes are ordered according to their inner products with it. For a certain choice of the gap-statistic, another gene cluster is created.

The idea behind the gap-statistic is that the size k of each cluster can be determined by minimizing a measure of within-group variation, that is, the relative size of the percentage of the total variance in a cluster, V_T, versus the variance between samples, V_B:

$$D_k = 100 \frac{V_B}{V_T}.$$

However, this percentage varies even when the rows and columns are independent, especially with small clusters, and so directly comparing these percentages may not be possible. Hastie et al. (2000) attempt to overcome this using multiple permutations of the data matrix, X. Each permutation data matrix, X^{*b}, $b = 1, \ldots, B$ is created by randomly reassigning all the expressions levels for a gene among the different samples, and doing this separately for all the genes. From each of the B X^{*b} matrices, equivalent coefficients of variation are calculated for all of the potential clusters.

The gap function is then defined as

$$Gap(k) = D_k - \overline{D_k^*},$$

where $\overline{D_k^*}$ is the average of these B permutation coefficients of variation and D_k is the original coefficient of variation for the cluster with k genes. More details about the practical implementation of the method can also be found in Chipman et al. (2003).

Model-based clustering

One of the drawbacks of all the previously discussed methods is the lack of explicit probabilistic justification. Many of the clustering methods are implementations of algorithms, rather than being informed by a general principle. An explicitly statistical way of dealing with clustering is by setting up a probabilistic model of how the unknown cluster labels c for the genes determine the observed data x_i, that is, for $c \in \{1, 2, \ldots, k\}$,

$$p(x_i \mid c). \tag{7.6}$$

By either maximizing this quantity, called the *likelihood*, or the *posterior* probability,

$$p(c \mid x_i) \propto p(x_i \mid c)\pi_c$$

over c, an explicitly probabilistic rule for assigning gene i to class c can be deduced.

Mixture models. A model whereby the likelihood of the data is specified relative to a discrete set of labels is called a mixture model. This comes from the fact that the unconditional distribution of the data, that is, before you know the true cluster label, is given as a mixture of the densities in Equation 7.6,

$$p(x_i) = \sum_{c=1}^{k} \pi_c p(x_i \mid c).$$

Mixture model approaches to clustering microarray data have been proposed by several people. Ghosh and Chinnaiyan (2002) assume that the data follow a multivariate normal distributed conditional on the class label c,

$$x_i \mid c \sim N(\mu_c, \Sigma_c).$$

It has been noted that by setting $\Sigma_c = \sigma^2 I$ for all c, mixture modelling reduces to k-means clustering (Venables and Ripley 1999). Ghosh and Chinnaiyan (2002) use the EM algorithm to maximize the likelihood in Equation 7.6 in order to find the 'most likely' class label for each gene c. This method also provides a way to estimate the number of clusters K for the data. By comparing Bayes factors (Kass and Raftery 1995),

$$B_{kl} = \frac{p(x \mid K = k)}{p(x \mid K = l)}, \tag{7.7}$$

one can gauge the relative evidence for a model with k clusters as compared to a model with l clusters.

Table 7.3 Six different variance–covariance structures for a multivariate normal distribution.

Σ	Distribution	Volume	Shape	Orientation
a.	Spherical	Equal		
b.	Spherical	Varying		
c.	Diagonal	Equal	Equal	
d.	Diagonal	Varying	Varying	
e.	Ellipsoidal	Equal	Equal	Equal
f.	Ellipsoidal	Varying	Varying	Varying

Bayes factors are useful in comparing non-nested models. Moreover, they do not rely on asymptotic theory. One problem of using Bayes factor in practice is that Equation 7.7 involves evaluating two integrals. Especially, for large data sets, as are common in microarray experiments, this is typically infeasible.

Ghosh and Chinnaiyan (2002) use the function Mclust from the R library mclust to implement the model of Fraley and Raftery (2002). This function evaluates the *Bayesian information criterion* (BIC) to compare across models,

$$\text{BIC}(k, \Sigma) = 2 \log p(x \mid \hat{\theta}, k, \Sigma) - \nu(k, \Sigma) \log n,$$

where k stands for the number of clusters; Σ stands for six different types of variance–covariance structures and $\nu(k, \Sigma)$ stands for the number of free parameters in a model with k clusters and covariance structure Σ; $\hat{\theta}$ represents the maximum likelihood estimate of the parameters in the model (k, Σ) and n is the number of observations. Mclust considers six different covariance structures, given in Table 7.3. They relate to a parameterization for multivariate normal distributions proposed by Banfield and Raftery (1993).

There are also other ways to compare across models. The BIC is the Bayesian version of the original *information criterion* (AIC) by Akaike (1973), given as

$$\text{AIC}(M) = 2 \log p(x \mid \hat{\theta}_M, M) - 2\nu(M).$$

Just like the BIC, it is evaluated for each model M of interest. The model M for which the AIC takes on a maximum is the preferred choice.

We apply the mixture model to the data from the breast cancer experiment. Figure 7.17 shows the BIC for different choices of covariance structures and different choices of the number of clusters. The BIC attains its maximum for three gene clusters, modelled as a multivariate normal distribution with a diagonal covariance structure and equal volume and shape across the clusters (Figure 7.18). From Table 7.4, it is clear that the final clustering is quite similar to what you would get by PAM clustering. Model-based clustering has the advantage of possessing principled ways to assess model choice, whereas PAM is much faster and still works for large data sets.

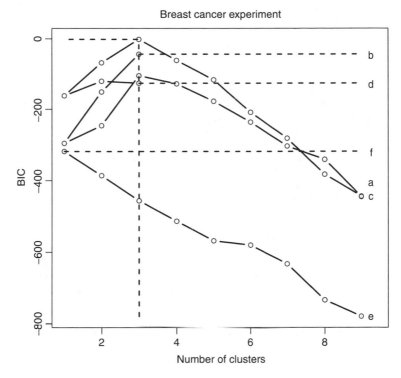

Figure 7.17 Bayesian information criterion evaluated for a variable number of gene clusters and over six different variance–covariance modes. The best model c attains a maximum for three clusters.

Table 7.4 Breast cancer experiment: strong consistency between classification via mixture modelling clustering and k-medoids (PAM), where $k = 3$.

	1	2	3
1	17	8	1
2	4	20	0
3	0	2	7

Other implementations of mixture models to gene clustering in microarray data can be found in Medvedovic and Sivaganesan (2002), who use Gibbs sampling to estimate infinite mixtures to obtain additional flexibility for unusual observations. McLachlan et al. (2002) is a careful study of applying mixture models to

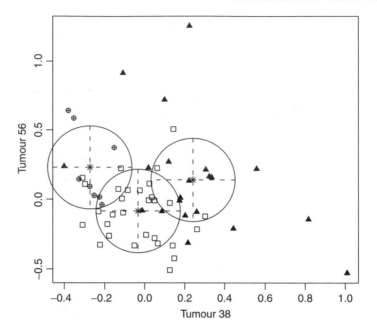

Figure 7.18   Superimposition of the three gene clusters found by mixture model clustering on the scatter plot of two tumours.

clustering samples, rather than genes. They have also produced software (EMMIX-GENE) that uses the EM algorithm to find the maximum likelihood solution to the clustering problem.

CAGED   Variations on the mixture model theme have recently been introduced to describe gene expression data with additional structure. Ramoni et al. (2002) propose a mixture model of autoregressive processes for time-course microarray experiments, which they have named *cluster analysis of gene expression dynamics* (CAGED).

In time-course experiments, a gene profile x_i for gene i consists of expression values over T time points,

$$x_i = (x_{i1}, x_{i2}, \dots, x_{iT}).$$

CAGED assumes that the gene expression profiles arise from k distinct dynamic processes. If gene i belongs to cluster $C(i) = c$, then the expressions x_{it} are assumed to be linear perturbations of the previous q values,

$$x_{it} = \beta_1^c x_{i,t-1} + \beta_2^c x_{i,t-2} + \dots + \beta_q^c x_{i,t-q} + \epsilon_{it},$$

where the parameters $\beta_1^c, \dots, \beta_q^c$ are common to all genes in cluster c. Effectively, this means that the time-course gene expressions are modelled according to an autoregressive process AR(q) or order q. The choice of the number of cluster k and the autoregressive order q can be handled in principle within the model itself. The aim is to maximize the posterior probability of the model M,

$$p(M|x) \propto p(x|M)p(M),$$

where $M = (k, q, C(1), \ldots, C(p), B, \sigma^2)$ are the model parameters, including the $k \times q$ matrix B of coefficients β_i^c. Assuming that the term $\epsilon_{it} \sim N(0, \sigma^2)$ and that gene profiles given their class labels are independent, Ramoni et al. (2002) show that $x|M$ is a multivariate normal with a tractable covariance structure.

Although, in principle, the problem is suitable for an MCMC (Monte Carlo Markov Chain) implementation, the space of all models $\mathcal{M}$ is too large for good mixing. The implementation of CAGED makes use of several search heuristics. The procedure starts by assuming that each of the p observed time series is generated by its own AR process with total likelihood $f(x|M_p)$. The next step merges two profiles together. This involves the computation of $p(p-1)/2$ likelihoods $f(x|M_{p-1}^j)$. The model with maximal marginal likelihood $f(x|M_{p-1}^*)$ is chosen as long as $f(x|M_{p-1}^*) > f(x|M_p)$. This procedure is repeated until merging reduces the marginal likelihood,

$$f(x|M_{k-1}^*) \leq f(x|M_k^*).$$

The model M_k^* is the final choice for the clustering model. For the actual algorithm to find the best marginal likelihood $f(y|M_i^*)$, CAGED uses heuristics to reduce the search space.

CAGED and other model-based clustering algorithms are part of an interesting development of introducing sound statistical techniques in a field that has been dogged by a lot of *ad hoc* choices. A related approach is due to Luan and Li (2003), who cluster time-course gene expression data using a mixed-effects model with B-splines. Despite our praise for model-based clustering, one should also keep in mind that if clustering is an exploratory technique, then there is certainly room for *ad hoc*ery. We hope to have struck a balance between what can be achieved with formal techniques and what might set the imagination alight.

8

Differential Expression

Differential expression indicates the changing of transcription levels across different phenotypes or conditions. The idea is that these transcription changes might be responsible for—or caused by—the change in phenotype. For example, the genes responsible for the presence of a certain disease X will be transcribed at a different rate than when the disease is absent. The following three pairs of terms will be used interchangeably in this chapter: (i) *differentially expressed* versus *not differentially expressed*, (ii) *active* versus *inactive* and (iii) *affected* versus *unaffected*.

8.1 Introduction

Many microarray experiments are aimed at finding 'active' genes. In the simplest case, one compares two different situations, for example, *treatment* versus *control* or two different conditions. The aims of such experiments are (i) to find which genes behave differently under different conditions and (ii) to determine a measure of confidence for this different behaviour for each gene.

In the following discussion, we shall focus on the comparison of two conditions only, which is by far the most common. A number of methods have been proposed for the analysis of differentially expressed genes. Dudoit et al. (2002) and Ge et al. (2003) suggest an analysis based on hypothesis testing, whereas Efron et al. (2000) and Efron et al. (2001) suggest an empirical Bayes-based approach. Both methods are perfectly sensible and often lead to very similar results. The basic philosophies behind these methods are, however, quite different.

8.1.1 Classical versus Bayesian hypothesis testing

For many problems in statistics there are two main schools of thought: the classical and the Bayesian school. Roughly speaking, the classical school believes

Statistics for Microarrays: Design, Analysis and Inference   E. Wit and J. McClure
© 2004 John Wiley & Sons, Ltd   ISBN: 0-470-84993-2

that parameters, such as a population mean, are fixed but unknown quantities and that only observations are truly random. The Bayesian school, on the other hand, assumes that after observing them the data can be considered fixed, whereas our knowledge about the parameters is really random.

This leads to different approaches when it comes to hypothesis testing. Both of them are interested in the parameter

$$\theta_g = \text{population mean difference in gene expression for gene } g.$$

In particular, both of them might like to know more about whether or not this parameter is zero, that is, whether or not gene g is differentially expressed. This knowledge corresponds to having information about the parameter v_g, where

$$v_g = 1_{\{\theta_g \neq 0\}},$$

where 1_A is the indicator function on set A. Classical statisticians tend to be very dualistic about v_g. It is either zero or it is not. This is not a matter of probability, but of the unknown reality. Bayesians are perfectly happy to describe their knowledge about v_g in terms of probabilities, as it is just another parameter.

The *classical hypothesis testing* approach for a microarray experiment tests for each gene g the hypothesis that that particular gene is not differentially expressed. This negative hypothesis is called the *null hypothesis*, $H_{g,0}$. The result of a hypothesis test is an absurdity probability, the so-called *p-value*. As we explain later, this quantity is closely related to the false positive rate (FPR). The lower this probability, the stronger the evidence against the null hypothesis. If the *p*-value is less than a certain cut-off, then H_0 is rejected in favour of the *alternative hypothesis*, $H_{g,1}$, under which gene g is differentially expressed. In some cases a further differentiation is made as those genes that are over-expressed (i.e. more active in the condition of interest than in the other condition) or as those that are under-expressed (i.e. less active in the condition of interest than in the other condition).

In the *Bayesian approach*, it is assumed that for each gene g there is an unobservable variable, v_g, that defines the gene's activation status. If a gene is not differentially expressed, then we define $v_g = 0$. The result of Bayesian inference is an update of our knowledge of v_g's value in terms of a *posterior probability*. This probability expresses the likelihood that $v_g = 0$. As we explain later, this quantity is closely related to the false discovery rate (FDR). Casella and Berger (2001) give a good introduction to Bayesian hypothesis testing.

These two approaches and their associated jargon may seem daunting to the uninitiated, but each of the two ways of looking at the problem has its own merits and they are complementary, rather than opposed. In fact, hypothesis testing results differ more within each school than between the schools. The main difference does not come from the particular testing philosophy but from the choice of *test statistic*.

Both classical and Bayesian hypothesis testing require a *differential expression score*, z_g, which is to be calculated for each gene from the microarray data. In the hypothesis testing setting, it is sometimes called the *test statistic*. This score is a summary of the information about the activity status of the gene. It can take a variety of forms, as discussed in Section 8.2.2. The differential expression score forms the basis of deciding whether or not a gene is differentially expressed.

8.1.2 Multiple testing 'problem'

The multiple testing 'problem' arises when more than one hypothesis is tested simultaneously. The larger the number of hypotheses, the more likely we are to find extreme differential expression scores, even if all the null hypotheses are in fact true. Clearly, if each hypothesis is rejected at some fixed posterior probability or fixed p-value and the number of hypotheses grows, then it becomes more and more likely that at least one null hypothesis will be falsely rejected.

Is this a problem? Well, that depends on your interests. If you want to control some form of per hypothesis error rate, such as the false discovery rate (FDR) or false positive rate (FPR), then clearly the number of hypotheses is, in principle, irrelevant. However, if you are interested in controlling some type of overall error rate, such as the familywise error rate (FWER), then this might be a problem.

The multiple testing 'problem' is therefore principally *a thorn in the eye of the beholder*. Nevertheless, the large number of hypotheses does focus attention quite poignantly on the types of error rates that one might wish to control. The rest of the chapter will focus on practical methods, some classical, some Bayesian, to control different types of error rates.

8.2 Classical Hypothesis Testing

Hypothesis testing is probably the most popular statistical method within many of the other sciences. It has several clear advantages. A hypothesis test is a standardized methodology with an easily interpretable either/or conclusion. Particularly, when testing for differential expression, this is very useful. It allows biologists to interpret the statistical output of the hypothesis tests by themselves, declaring the genes on their microarray 'active' or 'inactive,' based solely on the provided p-values.

However, statistical tests are laden with potential pitfalls. Failing to reject the alternative hypothesis does not exactly mean that one can confidently accept the null hypothesis. There exist different error rates one could choose to control. Misunderstanding the 'correct' error rate can have large consequences when testing many genes. This chapter partially aims to point out hypothesis testing pitfalls and to explain the possible error rates.

This section suggests several classical hypothesis tests for several different circumstances. Many of these tests are implemented in standard statistical software, which makes their application straightforward. Ge et al. (2003) and Dudoit et al. (2002) provide good reviews on hypothesis testing, both of which have also been motivated by microarray experiments.

8.2.1   What is a hypothesis test?

If for each gene it has to be decided whether or not that gene is 'active' or 'inactive', then one way of re-describing this problem is by expressing it as a series of two mutually exclusive 'hypotheses'. The null hypothesis H_0 typically states the status quo or default assumption, whereas the alternative hypothesis H_1 asserts the opposite. The role of these two hypotheses is not symmetric. In fact, classical hypothesis testing always assumes by default that the null hypothesis is true, *unless* enough evidence to the contrary has been found.

This means that for n genes we have n pairs of mutually exclusive hypotheses,

H_{g0}:   gene g is not differentially expressed
H_{g1}:   gene g is differentially expressed.

It can never be known for certain which of these hypotheses is really true. However, the data that come from a microarray experiment provide evidence to a greater or lesser extent in favour of one or the other. *Hypothesis testing* is a systematic method to summarize the evidence in the data in order to decide between the two possible hypotheses.

Hypothesis testing is not unlike a judicial trial. A defendant is presumed innocent unless proven guilty. A trial evaluates the evidence against innocence, that is, the null hypothesis. Only if there is overwhelming evidence, does the jury find the defendant guilty. In other words, only if there is sufficient evidence against the null hypothesis, will this hypothesis be rejected in favour of its alternative.

Hypothesis testing consists of four components: the hypotheses, the test statistic, the error rate and the decision rule to control that error rate.

Hypotheses

A hypothesis test begins with two opposite statements. One of these statements is called the *null hypothesis* H_0. This is generally the *status quo* situation. In the case of differential expression analysis, this is assuming that each gene is inactive. The second statement, the *alternative hypothesis* H_1, simply states the opposite of the null hypothesis, that is, that each gene is differentially expressed.

Standard statistical textbooks stress that hypotheses should be formulated in terms of *populations*. For instance, when testing the efficacy of a particular diet drug, the null hypothesis would read 'H_0: the mean weight loss in the population of all potential drug recipients is zero.'

It may not seem immediately obvious what the population is when testing for the differential expression of a particular gene g. The population is *not* the collection of all genes. It is hardly interesting to know what the whole population of genes does on average. The population consists of the expression difference for that particular gene across the two conditions in the population of similar subjects.

Test statistic

A test statistic is an efficient, univariate summary of the data used to evaluate the truth of the hypotheses. The kind of summary one chooses depends on the circumstances. A commonly used test statistic for comparing two sample tests is the difference between averages divided by the standard deviation, also called the t-statistic. The value of the test statistic is not easily interpretable. In fact, most of the mathematical details in hypothesis testing involve the test statistic.

Typically, the test statistic is chosen so that when H_0 is true, it has a certain known distribution. Sometimes, it is chosen so that the unconditional distribution can be estimated with bootstrap methods.

Error rates

When testing a single pair of hypotheses, there are two possible errors. In the case of an error, either the null hypothesis is wrongly rejected, a *false positive*, or wrongly accepted, a *false negative*. There is a trade-off between the probabilities of the two types of errors. By reducing the significance level and thereby reducing the probability of a false positive, one tends to reduce the so-called *power of a test*, increasing the probability of a false negative. The power is defined as one minus the false negative rate (FNR).

In a microarray setting, often thousands of tests are considered simultaneously— one for each gene on the array. In these kinds of situations, Table 8.1 describes the possible errors one can make while testing n genes. Most traditional methods focus on controlling FPR, that is, the expected fraction of false positives,

$$FPR = E\left[F_P/n_0\right].$$

The familywise error rate (FWER) is the probability that among all those genes that are inactive at least one is incorrectly classified as active,

$$FWER = P(F_P > 0).$$

Methods for controlling the FWER are often used in practice. However, the FWER is a very conservative error rate. Especially, with a large number of hypotheses, it

Table 8.1   Numbers of correct and incorrect conclusions of n hypothesis tests.

	Declared 'inactive'	Declared 'active'	Total
'Inactive'	T_N	F_P	n_0
'Active'	F_N	T_P	$n - n_0$
Total	$n - S$	S	n

is typically impractical to insist that the probability of making even only one false rejection should be small.

The false discovery rate (FDR) is in spirit closer to the FPR than the FWER. It is the expected number of inactive genes among those that are declared active,

$$FDR = E\left[F_P/S\right].$$

Methods controlling the FDR are relatively new and rely on some assumptions. However, they tend to be more suited for studies of an exploratory, rather than, confirmatory nature.

Decision rules

A decision rule is a method to translate the vector of n observed test statistics $z = (z_1, \ldots, z_n)$ into a set of n binary decisions, such that the error rate of choice is controlled at a preset level. As is shown in Section 8.2.3, all classical decision rules for any error rate use p-values to reject or accept a hypothesis.

The p-value itself controls the false positive rate (FPR). By comparing it to a pre-specified significance level α for a single hypothesis test, the probability of a false positive is less than α. For multiple hypothesis tests, the term F_P is a sum of indicator functions, $1_{\{p\text{-value}(i)\leq\alpha; v_i=0\}}$. Of these n functions, n_0 have an expectation less than α and the rest have an expectation of zero. Clearly, in that case $E(F_P/n_0) \leq \alpha$.

Any value may be chosen for the significance level. However, this cut-off needs to be decided prior to analysing the data. Deciding the cut-off *post hoc* makes hypothesis testing suspect as it effectively allows the experimenter to choose which hypotheses to reject.

***p*-value.** The p-value translates the value of the test statistic into a probability that expresses the *absurdity* of the null hypothesis. p-values close to zero indicate that the null hypothesis is *absurd* and should be rejected in favour of the alternative hypothesis. In the case of gene expression microarray experiments, small p-values mean that a gene is declared differentially expressed.

A confusing, but often heard, phrase is that a conclusion is *statistically signif-icant* if the p-value is close to zero. Although this is formally correct, we advise the biologists to use it sparingly because it is often confused with *physical signif-icance*. For example, in a very large study, a certain gene may be over-expressed in a *statistically significant* way by a mere 1.001 fold. Probably, this fold change is, biologically speaking, completely insignificant.

Many mistaken formulations of the p-value have entered the official literature. The p-value is *not* the probability that the null hypothesis is true. Instead, it represents how likely the observed data would be, if in fact the null hypothesis were true. More formally, the p-value is the probability of observing a value for the test statistic that is at least as extreme as the observed test statistic under the assumption that the null hypothesis is true.

The way the p-value is used is a bit like a *reductio ad absurdum* or a *proof by contradiction*. It assumes that the null hypothesis is true, and then reasons to a near absurdity. Then it concludes that the null hypothesis almost certainly has to be false.

Assumption	The null hypothesis is true.
Then	The probability that we observe the test T, that is, the p-value is (almost) zero.
But	We just did observe T.

Conclusion	The null hypothesis is (almost certainly) false.

The actual calculation of p-values tends to be quite mathematically involved. Fortunately, many statistical packages have routine ways to calculate them. For some unusual p-values, we have included the technical details in the appropriate sections.

8.2.2   Hypothesis tests for two conditions

Test statistics summarize the evidence in the data about the phenomena that are tested in the null hypothesis. Care must be taken to ensure that the statistic used is appropriate as not all statistics are appropriate for every situation. Different statistics give different p-values, and therefore using an inappropriate statistic will lead to inaccurate conclusions.

In this section, we discuss the main test statistics used in investigating differential expression. First, the commonly used t-statistic and variants of it are outlined. Then the robust Wilcoxon signed rank statistic is presented. Finally, we show how more complicated statistics can be used, such as the one due to Ideker et al. (2000), based on the likelihood ratio.

The Satterthwaite–Welch t-statistic

The t-statistic is perhaps the most popular statistic for testing the difference between two means. It is simple, and it has some optimality properties if the data are normally distributed.

When to use it?

- When the observations for each gene are independent.

- When there are no extreme outliers.

Depending on the data, different ways should be used to calculate the p-values:

- if there are a lot of observations (>30) in each sample, then use the

   Student t-distribution;

- if the observations are normally distributed, then use the **Student** t-**distribution**;

- if the samples are neither large nor are from a normal distribution, then use the **bootstrap**;

- if the samples differ only in their mean value parameter, then use the **permutation/randomization** method.

The statistic. The t-statistic is a standardized mean difference between the two samples for gene g:

$$z_g = \frac{\overline{x}_{g1\cdot} - \overline{x}_{g2\cdot}}{\sqrt{\dfrac{s_{g1}^2}{n_{g1}} + \dfrac{s_{g2}^2}{n_{g2}}}}, \tag{8.1}$$

where $\overline{x}_{g1\cdot}$ and $\overline{x}_{g2\cdot}$ are the means of the samples of expression values of gene g for the conditions 1 and 2, respectively; s_{g1}^2 and s_{g2}^2 are the sample variances of these two samples; n_{g1} and n_{g2} are the sizes of the two samples.

Essentially, Equation (8.1) scales the difference between the means of the two samples for the two conditions by a factor relating to the amount of variation in the two samples and to the size of these two samples. The more noise in the samples, the less clear the systematic difference between the two means will appear. The larger the sample, the smaller the impact of noise of each individual observation and therefore the clearer the systematic mean difference between the two samples.

Calculation of the p-values: Student t-distribution. The distribution of the t-statistic when there are different variances in each of the samples is unfortunately not exactly a Student t-distribution. Nevertheless, Satterthwaite (1946) found that under the null hypothesis of no differential expression the t-statistic is approximately t-distributed,

$$z_g \sim t_\nu,$$

where the degrees of freedom ν is given by

$$\nu = \frac{(\omega_1 + \omega_2)^2}{\omega_1^2/(n_1 - 1) + \omega_2^2/(n_2 - 1)}.$$

The quantities ω_i are the estimated squared standard errors $\omega_i = s_i^2/n_i$ for sample i. The p-value for the t-statistic is hence calculated by using the c.d.f. of the t_ν distribution:

$$p\text{-value for gene } g = 2 \times P\left(t_\nu \geq |z_g|\right).$$

The p-value is *twice* the exceedance probability, because this is a two-sided test in which there is no *logical* reason to expect over- or under-expression.

Calculation of the p-values: bootstrap. In order to be able to use the Student t-distribution to calculate the p-values, some strong assumptions have to be satisfied. Either the samples have to come from normal distributions, or the samples should be large enough for the central limit theorem to kick in. In the microarray setting, either of these two assumptions are unlikely to hold completely across all genes. Nevertheless, the normal approximation is, in many circumstances, an appropriate approximation.

The term *bootstrap* comes from the legendary Baron Munchausen who pulled himself out of a manhole by grabbing his own bootstraps. The statistical bootstrap (Efron 1979) attempts to do something similar by reusing the data many times over. It is a resampling method that simulates alternative values for the t-statistic in order to calculate an empirical p-value. The bootstrap works by creating a large number, B, of alternative values of the observed z_g, say $z_g^1, z_g^2, \ldots, z_g^B$, which resemble how z_g would be distributed if the null hypothesis were true. Then, an empirical p-value is calculated using this sample.

Each of these bootstrap values z_g^b is created by first resampling n_{g1} values from the observed values $x_{g11}, x_{g12}, \ldots, x_{g1n_{g1}}$ with replacement,

$$x_{g11}^b, x_{g12}^b, \ldots, x_{g1n_{g1}}^b,$$

then resampling n_{g2} values from $x_{g21}, x_{g22}, \ldots, x_{g2n_{g2}}$ with replacement,

$$x_{g21}^b, x_{g22}^b, \ldots, x_{g2n_{g2}}^b,$$

and finally calculating z_g^b according to a close relative of Equation (8.1),

$$z_g^b = \frac{(\overline{x}_{g1\cdot}^b - \overline{x}_{g2\cdot}^b) - (\overline{x}_{g1\cdot} - \overline{x}_{g2\cdot})}{\sqrt{\frac{(s_{g1}^b)^2}{n_{g1}} + \frac{(s_{g2}^b)^2}{n_{g2}}}}.$$

A large bootstrap sample $z_g^1, z_g^2, \ldots, z_g^B$ approximates the distribution of test statistic z_g under the null distribution that the gene g is not active. Notice, for example, that the sample is centered around zero.

The bootstrap sample is an approximation in two ways. First, the bootstrap sample forms only an empirical distribution. However, by making B large enough, this approximation can be made precise to any pre-specified level. Secondly, the accuracy of the bootstrap approximation is essentially limited by the number of observations n_{g1} and n_{g2} in the original samples. No resampling is going to make this approximation any more precise.

Hall and Martin (1988) propose what they call a *symmetric percentile t method* for estimating the p-value for the t-statistic with some excellent accuracy properties based on the bootstrap. We break the method down in some easily reproducible steps.

1. Create B bootstrap replicate statistics

$$z_g^b = \frac{(\overline{x}_{g1.}^b - \overline{x}_{g2.}^b) - (\overline{x}_{g1.} - \overline{x}_{g2.})}{\sqrt{\frac{s_{g1}^b}{n_{g1}} + \frac{s_{g2}^b}{n_{g2}}}}, \qquad \text{for } b = 1, \ldots, B, \qquad (8.2)$$

where $\overline{x}_{g1.}^b$ and $\overline{x}_{g2.}^b$ are the sample means and $(s_{g1}^b)^2$ and $(s_{g2}^b)^2$ are the sample variances of the bth bootstrap samples for condition 1 and 2.

2. Create the empirical distribution of the absolute values $|z_g^b|$ of the bootstrap sample. In particular, calculate the p-value from the empirical distribution of the absolute values at $|z_g|$,

$$p\text{-value for gene } g \approx \frac{\#\{|z_g^b| \geq |z_g|\}}{B},$$

where $\#\{|z_g^b| \geq |z_g|\}$ signifies the number of bootstrap replicate statistics greater than $|z_g|$.

This gives a useful test that can be used whenever there are sufficient replicates to create a bootstrap distribution. Simulations in Hall and Martin (1988) give some hope that appropriate p-values can be generated whenever the number of replicates for each gene in both conditions is in the order of 10. In that case, one would expect the p-values to be off by only 0.01 since the coverage error of the symmetric percentile t method is $O(n_{g1}^{-1} + n_{g2}^{-1})$.

By bootstrapping over all genes simultaneously, a joint null distribution that preserves the dependence structure can be created. This would allow the calculation of p-values that take into account high correlations among some genes.

Calculation of the p-values: permutation. Closely related to the bootstrap is the idea of *randomization* or *permutation* to calculate the p-values. The idea is to create a null distribution by randomly shuffling the observations from the two samples around many times. The resulting shuffled samples can then be used to get an empirical p-value, just like in the bootstrap situation.

The permutation method can only be used if the two distributions from which the two samples come are completely the same, except perhaps for the mean value parameter. In particular, if the variances of the two samples are not identical, the results of this method may be inaccurate.

The permutation method works by creating a large number, R, of alternative values of the observed value z_g, say $z_g^1, z_g^2, \ldots, z_g^R$, that resemble how z_g would be distributed if the null hypothesis were true. Then an empirical p-value is calculated using this sample.

Each of these bootstrap values z_g^r is created by first resampling n_{g1} values from the observed values $x_{g11}, \ldots, x_{g1n_{g1}}, x_{g21}, \ldots, x_{g2n_{g2}}$ *without* replacement,

$$x_{g11}^r, x_{g12}^r, \ldots, x_{g1n_{g1}}^r,$$

then assigning the remaining n_{g2} values to,

$$x^r_{g21}, x^r_{g22}, \ldots, x^r_{g2n_{g2}},$$

and finally calculating z^r_g according to Equation (8.1),

$$z^r_g = \frac{\overline{x}^r_{g1.} - \overline{x}^r_{g2.}}{\sqrt{\frac{(s^r_{g1})^2}{n_{g1}} + \frac{(s^r_{g2})^2}{n_{g2}}}}.$$

A large permutation sample $z^1_g, z^2_g, \ldots, z^R_g$ approximates the distribution of test statistic z_g under the null distribution that the gene g is not active. Notice, for example, that the sample is centred around zero. An approximate p-value can be defined as

$$p\text{-value for gene } g \approx \frac{\#\{|z^r_g| \geq |z_g|\}}{R}. \tag{8.3}$$

An example of two such resamplings is given in the box below. Note that only $\binom{n_{g1}+n_{g2}}{n_{g1}}$ different *combinations* of the results exist. If n_{g1} and n_{g2} are small, then R can be set to that number and a complete enumeration can be performed. However, small values also make the approximation in Equation (8.3) less accurate. As a rule of thumb, this permutation method is only recommended if both samples have at least 10 observations in it.

By permuting all genes in one microarray simultaneously, one is able to preserve the covariance structure of the data. This can be a good way to get around assuming independence of the genes.

This permutation/randomization method gets around having to make distributional assumptions about the data and as such is very useful for a microarray experiment. However, it only works under the assumption that both distributions are equal, except perhaps for a possible difference in their means. This can be a strong assumption, and serious consideration should always be given to using the bootstrap method instead.

Example 8.1 Creating randomization and bootstrap samples
In this example, we show how permutation and bootstrap samples can be obtained from an imaginary microarray experiment. The experiment has four biological replicates for each of two conditions. We focus, in the example, on the data for one gene:

Sample	Condition 1				Condition 2			
Original x_{gik}	−2.1	10.3	0.3	0.1	−0.3	0.2	−3.4	−1.8
(labels)	1	2	3	4	5	6	7	8

Examples of randomization/permutation samples:

Sample	Condition 1				Condition 2			
Samples 1	−1.8	0.1	10.3	0.2	−2.1	0.3	−3.4	−0.3
(combination)	8	4	2	6	1	3	7	5
Samples 2	−3.4	10.3	−0.3	−0.3	0.1	−2.1	−1.8	0.3
(combination)	7	2	5	6	4	1	8	3

Examples of bootstrap samples:

Sample	Condition 1				Condition 2			
Bootstrap samples 1	0.1	10.3	0.3	0.3	−0.3	−1.8	−3.4	−0.3
(resampling)	4	2	3	3	5	8	7	5
Bootstrap samples 2	10.3	−2.1	10.3	10.3	−0.3	0.2	−0.3	−3.4
(resampling)	2	1	2	2	5	6	5	7

Pooled variance t-statistic

There are a number of variants to the Welch t-statistic. One of them is the pooled variance t-statistic. As the name suggests, it pools the variances of the two samples to get a better estimate of the variance. However, it is only suitable if the variances of the two populations from which the samples come are approximately equal.

When to use it?

- When the population variances are (approximately) equal.

- When the observations for each gene are independent.

- When there are no extreme outliers.

Depending on the data, different ways should be used to calculate the p-values:

- if there are a lot of observations (>30) in each sample, then use the **Student t-distribution**;

- if the observations are normally distributed, then use the **Student t-distribution**;

- if the samples are neither large nor come from a normal distribution, then use the **bootstrap**;

- if the samples differ only in their mean value parameter, then use the **permutation/randomization** method.

A test of the equality of variances between the two populations can be done by comparing s_{g1}^2/s_{g2}^2 to the $F(n_{g1} - 1, n_{g2} - 1)$ distribution's function. If it is less than the 97.5 percentile and more than the 2.5 percentile, then we can conclude that the two populations have equal variances. Checking whether the variances of the two expression populations can be difficult for thousands of genes simultaneously. It makes good biological sense, however, that the same gene under different conditions should have the same variation. If the assumption of equal variances is particularly in doubt, the ordinary Welch t-statistic in Equation (8.1) can be used.

The test statistic.   The pooled variance t-statistic varies slightly from the one given in Equation (8.1) by pooling the information from both samples about the variance the two populations share. The statistic for gene g is given by

$$z_g = \frac{\overline{x}_{g1\cdot} - \overline{x}_{g2\cdot}}{\sqrt{\dfrac{s_{gp}^2}{n_{g1}} + \dfrac{s_{gp}^2}{n_{g2}}}}, \tag{8.4}$$

where

$$s_{gp}^2 = \frac{(n_{g1} - 1)s_{g1}^2 + (n_{g1} - 1)s_{g2}^2}{n_{g1} + n_{g2} - 2}$$

is the pooled sample variance for gene g.

Calculation of the p-value.   For the calculation of the p-value, similar considerations apply as with the Satterthwaite–Welch t-statistic given above. However, if one opts to calculate the p value with the t-distribution, then the degrees of freedom are much simplified. Under the null hypothesis of no differential expression,

$$z_g \sim t_{n_{g1}+n_{g2}-2},$$

where $t(n_{g1} + n_{g2} - 2)$ is the t-distribution with $n_{g1} + n_{g2} - 2$ degrees of freedom. The p-value for the t-statistic is hence calculated by using the c.d.f. of the $t_{n_{g1}+n_{g2}-2}$ distribution.

$$p\text{-value for gene } g = 2 \times P\left(t_{n_{g1}+n_{g2}-2} \geq |z_g|\right).$$

The p-value is *twice* the exceedance probability, because this is a two-sided test in which there is no *logical* reason to expect over- or under-expression.

Wilcoxon rank sum statistic

The Wilcoxon rank sum statistic, also called the Mann–Whitney statistic, is a *nonparametric* test statistic. It uses the ranks of observations, rather than the values

themselves, to compare the two samples. This makes it more *robust* than the
t-statistic, in the sense that it is not very sensitive to outliers.

When to use it?

- When the x_{gik} do not appear to be normally distributed. Otherwise, it will
  be better to use the t-statistic, which has greater power than the Wilcoxon
  test when the data are normally distributed.

- When the distributions of the two samples have approximately the same
  general shape.

- When there are at least four replicates for the two conditions and preferably
  six or more. Having fewer replicates than four will make it impossible for
  any data set to have a p-value less than 0.05.

Although it is a formal requirement that the general shape of the two distributions is approximately the same, the Wilcoxon rank sum test is quite robust against slight deviations.

The test statistic. The Wilcoxon differential expression score is calculated by ranking all the $n_{g1} + n_{g2}$ values from both samples and then summing up the ranks associated with the one of the two samples. Formally, the Wilcoxon rank sum statistic is defined as

$$z_g = \sum_k \text{Rank}(x_{g1k}),$$

where $\text{Rank}(x_{g1k})$ is the rank of x_{g1k} amongst all the x_{g1k} and x_{g2k} values, ranging from 1 for the smallest value and $n_{g1} + n_{g2}$ for the highest value. If all the x_{g1k} values are smaller than all the x_{g2k} values, then z_g will be $\sum_{k=1}^{n_{g1}} k$. If all the x_{g1k} values are larger than all the x_{g2k} values, then z_g will be $\sum_{k=n_{g2}+1}^{n_{g1}+n_{g2}} k$. All other possibilities lie between these two values.

Calculation of the p-value. Under the null hypothesis, the statistic will simply be sum of a randomly chosen combination of n_{g1} numbers from the range 1 to $n_{g1} + n_{g2}$. Here the distribution of the statistic is

$$Z_g \sim \text{Wilcoxon}(n_{g1}, n_{g2}).$$

We can calculate a p-value for the test statistic from the tables of the Wilcoxon distribution. When observing a value z_g, the p-value is twice the probability that a $\text{Wilcox}(n_{g1}, n_{g2})$ achieves such an extreme value, that is,

$$p\text{-value for gene } g = 2 \times \min\left\{P(Z_g \leq z_g), P(Z_g \geq z_g)\right\}.$$

Global error likelihood ratio test statistic

Another method is the global error likelihood ratio test statistic put forward by Ideker et al. (2000). It assumes that the errors in each sample have constant variance over all of the genes. The method uses likelihood ratios to form the test statistic.

The two likelihoods considered in the ratio are the one for the null hypothesis model and the one for the alternative hypothesis model. The *likelihood* in statistics is the probability of getting the data values we find when a particular model is assumed to be true.

When to use it? For this test statistic the expression values should be unlogged; however, they can be background subtracted and adjusted for array and dye effects. The statistic assumes that the variation in the lower expression range is mainly additive, whereas in the higher ranges it is mainly multiplicative.

The test statistic. The intuitive idea of this statistic is to look for each gene at the ratio of the likelihood of differential expression versus no differential expression,

$$\lambda_g = \frac{\text{likelihood of differential expression for gene } g}{\text{likelihood of no differential expression for gene } g}.$$

If this ratio is large, it suggests that there is quite some evidence for the case that gene g is differential expressed. However, the actual calculation of the test statistic, let alone the calculation of the p-value, is not straightforward.

Calculating the test statistic: mathematical details. Normally, a test statistic can be calculated directly on the basis of the data. This test statistic is slightly different in that respect. For each gene g, λ_g is *estimated* on the basis of all expression data $\{x_{gij}\}$ simultaneously. Then, assuming that there is no differential expression for gene g, the p-value is calculated.

The test statistic involves a ratio of likelihoods of the data under the null and alternative hypothesis. The data are assumed to be generated by an error model that includes both multiplicative and additive elements:

$$\begin{aligned} x_{g1k} &= \mu_{g1} + \mu_{g1}\varepsilon_{g1k} + \delta_{g1k} \\ x_{g2k} &= \mu_{g2} + \mu_{g2}\varepsilon_{g2k} + \delta_{g2k} \end{aligned} \tag{8.5}$$

where x_{g1k} and x_{g1k} are the expression levels under condition 1 and 2 respectively, for the kth replicate of gene g; μ_{g1} and μ_{g2} are the true mean expression of gene g in two conditions; ε_{g1k} and ε_{g2k} are the multiplicative error terms and these have a bivariate normal distribution with mean 0, respective variances $\sigma_{\varepsilon1}^2$ and $\sigma_{\varepsilon2}^2$, as well as a correlation ρ_ε; δ_{g1k} and δ_{g2k} are the additive error terms and these have a bivariate normal distribution with mean 0, respective variances $\sigma_{\delta1}^2$ and $\sigma_{\delta2}^2$, as well as a correlation ρ_δ.

In addition to modelling the error in terms of both multiplicative and additive terms, the model allows the errors for the two conditions to be correlated (via ρ_ε and ρ_δ).

The parameters $\beta = (\sigma_{\varepsilon_1}, \sigma_{\varepsilon_2}, \rho_\varepsilon, \sigma_{\delta_1}, \sigma_{\delta_2}, \rho_\delta)$ and $\mu = \{(\mu_{g1}, g2) : g = 1, \ldots, n\}$ are estimated using a maximum likelihood approach, optimizing

$$
L(\beta, \mu) = \prod_{g=1}^{n} L_g(\beta, \mu_{g1}, \mu_{g2})
$$

$$
= \prod_{g=1}^{n} \prod_{k=1}^{K} p(x_{g1k}, x_{g2k} | \beta, \mu_{g1}, \mu_{g2}).
$$

The model as it stands allows the means for the two conditions to be different. This makes it suitable for describing the alternative hypothesis situation—where the gene is differentially expressed in the two conditions.

A second maximization occurs using the same likelihoods constrained so that $\mu_{g1} = \mu_{g2} = \mu_g$. This defines the situation under the null hypothesis.

It is then possible to use the following log ratio as our test statistic:

$$
\lambda_g = -2 \ln \left(\frac{\max_{\mu_g} L_g(\beta, \mu_g, \mu_g)}{\max_{\mu_{g1}, g2} L_g(\beta, \mu_{g1}, \mu_{g2})} \right).
\tag{8.6}
$$

When the null hypothesis is true, then the ratio in Equation (8.6) will have an expected value of 1. If the gene is differentially expressed, the ratio will have an expected value below 1 and so λ_g will have a larger expected value than in the null case.

Calculation of p-values. Under the null hypothesis, where $\mu_{g1} = \mu_{g2} = \mu_g$, the distribution of Equation (8.6) will follow a χ^2 distribution with 1 degree of freedom. One minus the inverse c.d.f. of the χ^2 distribution will then give the p-value. Note that here we are only interested in the upper tail of the null distribution, that is, where the gene is active. The model where μ_{g1} and μ_{g2} are allowed to vary will be expected to fit the data better and hence give a larger likelihood; this will lead to large values of λ_g.

For this reason, only the probability obtained from 1 minus the c.d.f. is considered, and it is not doubled in order to create the p-value, unlike the situation for the t and Wilcoxon statistics.

8.2.3 Decision rules

Which decision rule should be used depends on what error rate one wants to control. This depends, first of all, on whether one is interested in only one gene or more than one gene. Further, if one is interested in multiple genes, there are several possible error rates to choose from.

One gene of interest

The decision rule chosen depends crucially on how many genes we are truly interested in. When only one gene is of interest in a microarray experiment, the decision rule can use the individual, unadjusted p-values to control the false positive rate (FPR).

In other words, we should disregard all the data on other genes. This may sound extreme and wasteful but, when the activity of other genes is not of interest, testing the other genes is unnecessary and provides no further information. It can also encourage the use of multiple testing correction procedures that are wholly inappropriate when it is only one test we are interested in.

Multiple genes of interest

When testing multiple genes, one could still choose to control the FPR. Other error rates that might be of interest are the familywise error rate (FWER) and the false discovery rate (FDR). Which one is suitable depends on the question we wish to answer. If one is specifically interested in the question of whether the genes that our procedure classifies as 'active' are *all* actually active or whether at least one of them is not, then a procedure controlling FWER should be employed. If one does not mind the presence of some false rejected hypothesis, then a FDR or a FPR controlling procedure should be used.

All of the decision rules make use of the p-values, $P_1, \ldots, P_n$, of the individual hypotheses. In each of these methods, we shall need the order statistics of the p-values. Let

$$P_{(i)} = i\text{th smallest } p\text{-value}$$

amongst all the p-values calculated for each of the hypotheses.

Conservative error rate: familywise error rate. A procedure that controls the FWER allows the user to control the probability that at least one gene will be falsely declared significant among the set of genes considered. Such procedures have long been the most commonly used ones for dealing with the problem of multiple testing. Here, two of the methods for controlling FWER are described— the Bonferroni correction and Hochberg's procedure (Hochberg 1988).

The size of the sets of genes rejected by these two procedures can vary greatly; however, both methods do control the familywise error rate. The difference is that the methods have different levels of *power*. Power is the expected proportion of truly active genes that are correctly identified as being active; it is therefore generally preferable to have a method with greater power. The most extreme example of a FWER controlling procedure is the Bonferroni correction. It has the advantage of being extremely simple to implement. Unfortunately, this comes at the cost of reduced power.

Bonferroni FWER Procedure. *Classify all genes that have an associated p-value less than α/n as active. This procedure guarantees that the FWER is less or equal to α.*

The Bonferroni method simply divides the desired FWER, α, by the number of hypothesis tests being conducted, that is, the number of genes being considered and uses this as the rejection cut-off for each of the individual p-values. To control the FWER at a 0.05 level with 1,000 comparisons, a rejection cut-off of 0.00005 for each individual gene's p-value is required. Though Bonferroni controls the FWER very well, it has very low power—it will fail to reject many truly differently expressed genes.

Bonferroni, and similar procedures such as Šidák's method (Dudoit et al. 2002), are single-step FWER control methods—ones where the p-values are all tested against the same cut-off level (e.g. α/n in Bonferroni FWER procedure). These methods are more extreme and have less power than the other main group of FWER controlling procedures—*step-down* ones. Step-down methods still control FWER but gain power by only subjecting the smallest p-value, $P_{(1)}$, to the single-step level test; larger p-values are subjected to progressively less stringent bounds. Hochberg's procedure is a simple step-down method for controlling FWER.

Hochberg's FWER Procedure. *Let k be the largest g $(0 \leq g \leq n)$ for which*

$$P_{(g)} \leq \frac{\alpha}{n - g + 1};$$

then reject all $H_{(g)}$, for $g = 1, 2, \ldots, k$, where $H_{(g)}$ is the null hypothesis associated with the gene with the gth smallest p-value. This procedure guarantees that the FWER is less or equal to α.

This means that if the largest p-value is less than α, $P_{(n)} \leq \alpha$, then all hypotheses are rejected. If this is not the case, then $H_{(n)}$ cannot be rejected and one goes on to compare the second largest p-value $P_{(n-1)}$ with $\alpha/2$. If $P_{(n-1)} \leq \alpha/2$, then all $H_{(i)}$ $(i = 1, \ldots, n - 1)$ are rejected. If not, then $H_{(n-1)}$ cannot be rejected and one proceeds to compare $P_{(n-2)}$ with $\alpha/3$, and so on.

Dudoit et al. (2002) suggest a FWER control proposed by Holm (1979). However, this procedure has been superseded by Hochberg's sharper procedure that has greater power. Although the Hochberg FWER procedure has greater power than the Bonferroni correction, it might not lead to any more genes being declared active in many practical microarray settings. The step-down Hochberg procedure is little different from the Bonferroni correction when g is small and n is large as in that case

$$\frac{\alpha}{n - g + 1} \approx \frac{\alpha}{n}.$$

Procedures for controlling FWER have been used in the analysis of microarrays and will be appropriate if it is this type of error that the experimenter needs to control. Often, however, the low power these methods have for detecting truly active genes means that controlling other errors, like the FDR, is preferred.

Finally, note that both the Bonferroni correction and the Hochberg FWER procedure have *strong control* of the FWER; this means that they control FWER whatever the proportion of truly differently expressed genes. *Weak control* on the other hand means that the procedure only controls FWER when none of the genes is differentially expressed.

Less stringent error rate: false discovery rate. The definition of the false discovery rate of a set of multiple tests is the average proportion of inactive genes among those that were declared active. A FDR procedure allows the user to control this expectation. For example, by deciding to accept a false discovery rate of 5% for a microarray experiment, a FDR procedure will find the largest subset of genes to be classed as differentially expressed that has an expected percentage of inactive genes of 5%.

Benjamini and Hochberg (1995) propose a procedure to control the FDR. It is a *step-down* procedure.

Benjamini and Hochberg FDR procedure. *Let k be the largest g $(0 \leq g \leq n)$ for which*

$$P_{(g)} \leq \frac{g\alpha}{np_0};$$

then reject all $H_{(g)}$, for $g = 1, 2, \ldots, k$, where $H_{(g)}$ is the associated null hypothesis. This guarantees that the FDR is less than or equal to α.

Unfortunately, to implement the procedure precisely, the true fraction p_0 of 'inactive' genes is required. Obviously, this fraction will not be known *a priori*. Replacing p_0 by 1 will guarantee that the FDR is controlled conservatively. Under more stringent assumptions, such as those we impose in Section 8.3.2, it is possible to estimate p_0.

8.2.4   Results from skin cancer experiment

In the skin cancer experiment (Section 1.2.2), gene expression in normal skin is compared to gene expression in cancerous skin. Here, we use the skin cancer data to compare the results from different error rate controls and test statistics.

For these data, t-statistics and Wilcoxon statistics are calculated for each gene, along with their p-values; the different procedures for creating decision rules are then applied to these p-values, and the numbers of genes rejected by each procedure and statistic combination are compared. Table 8.2 shows the number of genes declared 'active' for each decision rule when the different error rates are all controlled at the 5% level.

When the decision procedure rejects H_0 if the unaltered p-values are smaller than $\alpha = 0.05$, then we control the false positive rate (FPR). This FPR procedure

Table 8.2   Comparison of error rates and test statistics in the skin cancer experiment.

Decision rule	t	Wilcoxon
FPR	507	494
FWER (Bonferroni)	14	0
FWER (Hochberg)	14	0
FDR (Benjamini and Hochberg)	91	67

declares that around 500 out of the 4,608 genes tested are differentially expressed. Even if none of the genes are in fact differentially expressed, we would expect around 230 of them to be classed 'active'.

Using either of the two FWER controlling procedures (the Bonferroni and Hochberg) leads to very few genes (14) being classed as 'active' when using the t-statistic and none when the Wilcoxon statistic is used. Though it is unlikely that we have falsely classed any inactive genes as 'differentially expressed' here, it is likely that we have overlooked many truly active genes.

The Benjamini and Hochberg's FDR procedure selects between 50 and 100 genes. This procedure has an expected proportion of falsely classified genes among the declared 'active' genes of 5%. This means, roughly speaking, that of the 91 genes selected as 'active' with the t-statistic, we expect that around 5 will be, in fact, 'inactive'; of the 67 chosen using the Wilcoxon statistic, we expect around 3 to be actually inactive. For a more precise interpretation of the FDR, see Section 10.1.3.

The Wilcoxon statistic is more robust than the t-statistic if the data are few and not normally distributed. It is our experience, however, that appropriately normalized and log-transformed gene expressions typically do not deviate far from a normal distribution. Especially, given the exploratory rather than confirmatory nature of microarray analysis in general, we believe that making parametric assumptions is a good idea.

The t-statistic and Wilcoxon methods described here do not assume a common variance for each of the genes—they consider each gene separately. The likelihood ratio test described in Section 8.2.2 gains some power by making the additional assumption of a common variance model.

8.3   Bayesian Hypothesis Testing

In this section, we define Bayesian hypothesis testing as an alternative to classical hypothesis testing. We shall follow the same familiar hypothesis testing structure as before, except that we shall define an alternative to the classical p-values.

Hypotheses

Since the purpose of differential expression analysis is to find genes that have different mean DNA transcription levels across two conditions, the parameters of interest are the hidden variables that determine whether each gene is differentially expressed or not:

$$v_g = \begin{cases} 1 & \text{if gene } g \text{ is differentially expressed} \\ 0 & \text{otherwise.} \end{cases} \tag{8.7}$$

In Bayesian hypothesis testing, the variable v_g takes over the role that the null and alternative hypotheses, H_{g0} and H_{g1}, play in classical hypothesis testing. Unlike in classical hypothesis testing, however, there is no asymmetry in the role of $v_g = 0$ and $v_g = 1$.

Test statistics

The parameter v_g is not observable directly, and the data give partial information about it. Just as in classical hypothesis testing, we summarize the evidence in favour of differential expression into a *differential expression score*, that is, test statistic, z_g for each gene. Many choices for z_g are possible as long as this number expresses the observed *relative* activity for each gene.

Two of the statistics that are recommended are the common t-statistic and the non-parametric Wilcoxon rank sum statistic. Section 8.2.2 contains a number of other sensible two-sample statistics. The t-**statistic** expresses the amount of variation between the two conditions (numerator) relative to their respective levels of variation (denominator):

$$z_g = \frac{\overline{x}_{g1.} - \overline{x}_{g2.}}{se}, \tag{8.8}$$

where x_{gik} are normalized values of the kth replicate of the gth gene for condition i and se is the standard error. It is calculated as

$$se = \sqrt{s_{g1}^2/n_{g1} + s_{g2}^2/n_{g2}},$$

where s_{gi} are the estimated standard deviations among the observations for gene g under condition i. The quantity n_{gi} is the number of replicates for gene g under condition i. By normalizing this quantity with respect to the standard error, small but consistent fold changes are taken to be as important as large but variable fold changes. Sometimes it is possible to use the *pooled* standard error instead of the normal standard error,

$$se_p = \sqrt{\frac{(n_{g1} - 1)s_{g1}^2 + (n_{g2} - 1)s_{g2}^2}{n_{g1} + n_{g2} - 2}\left(\frac{1}{n_{g1}} + \frac{1}{n_{g2}}\right)},$$

if the assumption of equal variances across the two conditions (although not necessarily across the genes) is reasonable.

The **Wilcoxon rank sum statistic** (also called the Mann–Whitney statistic) is a more *robust* statistic, in the sense that it is less sensitive to outliers than the

t-statistic. It expresses the relative location of the replicates from the one condition as compared to the replicates of the other condition:

$$z_g = \sum_k \text{Rank}(x_{g1k}), \qquad (8.9)$$

where $\text{Rank}(x_{g1k})$ is the rank of observation x_{g1k} for gene g under the first condition among all $n_{g1} + n_{g2}$ observations across both conditions. When the two conditions have only few observations, the information contained in the Wilcoxon statistic is next to nothing. We recommend using the Wilcoxon rank sum statistic only when each of the two conditions has at least five, but preferably more, observations in it.

Example 8.2 Differential expression scores

For a certain gene g, the gene expression values x_{g1k} under condition 1 are given as $-2.1, 10.3, 0.3$ and 0.1. For the same gene, the values under condition two are $-0.3, 0.2, -3.4$ and -1.8. How can one calculate z_g for different choices of the test statistic?

For the t-statistic,

$$\bar{x}_{g1.} = \frac{-2.1 + 10.3 + 0.3 + 0.1}{4} = 2.2$$

$$s_{g1}^2 = \frac{(-2.1 - 2.2)^2 + (10.3 - 2.2)^2 + (0.3 - 2.2)^2 + (0.1 - 2.2)^2}{4 - 1} = 30.7$$

$$\bar{x}_{g2.} = \frac{-0.3 + 0.2 - 3.4 - 1.8}{4} = -1.3$$

$$s_{g2}^2 = \frac{(-0.3 + 1.3)^2 + (0.2 + 1.3)^2 + (-3.4 + 1.3)^2 + (-1.8 + 1.3)^2}{4 - 1} = 2.6$$

$$se = \sqrt{\frac{30.7}{4} + \frac{2.6}{4}} = 2.9$$

$$z_g = \frac{2.2 - (-1.3)}{2.9} = 1.2$$

For the Wilcoxon rank sum statistic,

Ordered expression	−3.4	−2.1	−1.8	−0.3	0.1	0.2	0.3	10.3
Condition	2	1	2	2	1	2	1	1
Rank	1	2	3	4	5	6	7	8

Then the Wilcoxon rank sum statistic is the sum of the ranks associated with condition 1,

$$z_g = 2 + 5 + 7 + 8 = 22$$

Note that the maximum possible value for z_g is 26, and the minimum possible value is 10.

Decision rule and error rate

Bayesian hypothesis testing attaches a probability to the statement 'gene g is not differentially expressed' when a certain large value z is found for the test statistic z_g. There are two main choices for defining this probability. Either we can consider the posterior probability of no differential expression when the test statistic is exactly equal to z,

$$p(v_g = 0 | z_g = z),$$

or we can consider the probability of no differential expression when the test statistic z_g exceeds z,

$$p(v_g = 0 | z_g \geq z). \tag{8.10}$$

The latter expression, Equation (8.10), is more in the spirit of classical hypothesis testing, where the p-value is defined as the exceedance probability $p(z_g \geq z | v_g = 0)$. In fact, we also prefer Equation (8.10) from the point of view of controlling the error rate. There is a direct link between the posterior 'inactivity' probability and the false discovery rate (FDR). This connection is made explicit by the following procedure.

Bayesian FDR procedure. (Wit et al. 2004b) *Let $\mathcal{R}$ be the set of genes whose test statistics exceed a certain cut-off z^*,*

$$\mathcal{R} = \{g | z_g > z^*\},$$

where the cut-off z^* is defined as,

$$z^* = \min\{z \mid p(v_g = 0 | z_g \geq z) = \alpha\}. \tag{8.11}$$

If we declare all the genes in R as differentially expressed, then the FDR is controlled at level α, that is,

$$FDR(\mathcal{R}) \leq \alpha.$$

Note that a similar cut-off z_*, under which genes are declared differentially expressed, can be found via

$$z_* = \max\{z \mid p(v_g = 0 | z_g \leq z_*) = \alpha\}.$$

For presentational simplicity, we only focus on the upper cut-off z^*

Related observations can be found in Efron et al. (2001) and Genovese and Wasserman (2002). It is closely related to the q-value (Storey 2002). The proof for this interesting property is very elementary.

$$FDR(\mathcal{R}) = E\left[\frac{F_P}{S} 1_{\{S>0\}}\right]$$

$$= E\left[\frac{\sum_{i=1}^{n} 1_{\{z_i > z^*; \, v_i = 0\}}}{\sum_{i=1}^{n} 1_{\{z_i > z^*\}}} 1_{\{S>0\}}\right]$$

$$= E \left[E \frac{\sum_{i=1}^{n} 1_{\{z_i > z^*; \ v_i = 0\}}}{\sum_{i=1}^{n} 1_{\{z_i > z^*\}}} 1_{\{S > 0\}} \mid 1_{\{z_1 > z^*\}}, \ldots, 1_{\{z_n > z^*\}} \right]$$

$$= E \left[\frac{\sum_{i=1}^{n} P(v_{i=0} \mid z_i > z^*) 1_{\{z_i > z^*\}}}{\sum_{i=1}^{n} 1_{\{z_i > z^*\}}} 1_{\{S > 0\}} \right]$$

$$= \alpha P(S > 0)$$

$$\leq \alpha$$

This means that in the unlikely case that the posterior 'inactivity' probability $p(v_g = 0 \mid z_g \geq z)$ is known, then the FDR can be controlled in a straightforward manner via the Bayesian FDR procedure in Equation (8.11).

The posterior probability can be broken down into three components using Bayes theorem,

$$p(v_g = 0 | z_g \geq z) = p_0 \frac{1 - F_0(z)}{1 - F(z)}.$$

Knowing the posterior probability is therefore equivalent to knowing,

- the fraction p_0 of 'inactive' genes on the array;

- the distribution F_0 of the test statistic under the null hypothesis, $v = 0$;

- the distribution F of the test statistic in the population, that is, the microarray.

As in classical hypothesis testing, the null distribution F_0 is typically known, such as for the t-statistic or the Wilcoxon statistic. However, the distribution F is some sort of mixture of the null distribution and the distribution from the 'active' genes, and is therefore not known. The quantity p_0 is typically also unknown.

In the following sections, we describe two methods to estimate F and p_0. Section 8.3.1 describes a completely general approach, among which we illustrate using the Wilcoxon rank sum test. In Section 8.3.2, we deal with specific estimation procedures for t-statistics.

8.3.1 A general testing procedure

This section describes a practical implementation of the general Bayesian FDR procedure. It turns out that the method, under certain circumstances, is equivalent to the classical FDR procedure by Benjamini and Hochberg (1995), described in Section 8.2.3.

Our aim is to estimate F and p_0 so as to obtain the function

$$\varphi(z) = \hat{p}(v_g = 0 | z_g \geq z)$$

$$= \hat{p}_0 \frac{1 - F_0(z)}{1 - \hat{F}(z)}.$$

If we want to control the FDR at level α, we then estimate the true rejection threshold z^* in Equation (8.11) via

$$\hat{z}^* = \min\{z \mid \varphi(z) = \alpha\}.$$

Let z be an arbitrary two-sample test statistic calculated for p genes, for which the distribution under the null hypothesis is known, that is,

$$z_g | \{v_g = 0\} \sim F_0, g = 1, \ldots, p.$$

For example, if z_g is the t-statistic, then we know that it has a t distribution with $n_1 + n_2 - 2$ degrees of freedom

$$z_g | \{v_g = 0\} \sim t_{n_1 + n_2 - 2},$$

where n_1 and n_2 are the number of observations of the first and second condition respectively. If z_g is the Wilcoxon rank sum statistic, then it has the Wilcoxon distribution with parameters n_1 and n_2 respectively,

$$z_g | \{v_g = 0\} \sim \text{Wilcoxon}(n_1, n_2).$$

The distribution for F is unknown because both the distribution under the alternative and the number of genes for which the alternative is true are unknown. The simplest estimate for F is the empirical distribution function of the data, that is,

$$\hat{F}(z) = \frac{1}{p} \sum_{g=1}^{p} 1_{\{z_g \leq z\}}. \tag{8.12}$$

By replacing p_0 with 1, we ensure that the estimate $\hat{p}(v_g = 0 | z_g \geq z)$ is, probabilistically speaking, conservative.

The posterior 'inactivity' probability can then be estimated as

$$\hat{p}(v_g = 0 | z_g \geq z) = \frac{1 - F_0(z)}{1 - \hat{F}(z)}, \tag{8.13}$$

where $\hat{F}$ is the empirical distribution function in Equation (8.12).

Theorem 8.1 Equivalence with Benjamini and Hochberg FDR procedure. *If $\hat{p}(v_g = 0 | z_g \geq z)$ is monotone in z, then the Bayesian FDR procedure using Equation (8.13) is equivalent with the Benjamini and Hochberg FDR procedure (Section 8.2.3).*

It is easy to show that for any choice of α, the set of rejected null hypotheses $\mathcal{R}$ in both instances is the same. First, we note that the ith smallest p-value $P_{(i)}$ is the exceedance probability of the $(n - i)$th order statistic $z_{(n-i)}$ under the null hypothesis,

$$P_{(i)} = 1 - F_0(z_{(n-i)}).$$

Because of the monotonicity of $\hat{p}(v_g = 0 | z_g \geq z)$ and Equation (8.12), the set of rejected hypothesis under the Bayesian FDR procedure $\mathcal{R}_{\text{Bayes}}$ is identical to the set of rejected hypothesis under the Benjamini and Hochberg FDR procedure $\mathcal{R}_{\text{BH}}$.

$$
\begin{aligned}
\mathcal{R}_{\text{Bayes}} &= \{g | \hat{p}(v_{(g)} = 0 | z_{(n-g)} \geq z) \leq \alpha\} \\
&= \{g | 1 - F_0(z_{(n-g)}) \leq \alpha(1 - F(z_{(n-g)}))\} \\
&= \{g | P_{(g)} \leq \alpha \frac{g}{n}\} \\
&= \mathcal{R}_{\text{BH}}.
\end{aligned}
$$

Example: Bayesian Wilcoxon rank sum test

The Wilcoxon rank sum test sums the ranks of the observations from one condition, as defined in Equation (8.9). If the replicate intensity levels for condition 1 are large compared to those of the other condition, then the Wilcoxon statistic will be large, and *vice versa*. The range of possible values for z_g is determined by the discrete number of replicates of each of the conditions in the experiments.

For example, for 8 replicates in each condition, the Wilcoxon rank sum statistic ranges between 36 (when all the condition 1 replicates are smaller than all of the condition 2 replicates) to 100 (where all the condition 1 replicates are larger than all the condition 2 replicates). If none of the genes were differentially expressed in an experiment, then the ranks of condition 1 would be 8 randomly chosen numbers from the range 1 to 16. The statistic is then said to follow the Wilcoxon(8,8) distribution, which is given by the solid line in Figure 8.1. An advantage of using the Wilcoxon statistic is thus that the null distribution of the test statistic is known and so does not have to be estimated from the data.

If at least some of the genes in a microarray experiment are differentially expressed, then this increases the weight of the tails. The points in Figure 8.1 are the empirical distribution of an actual microarray experiment and suggest that in this experiment there are more genes with extreme Wilcoxon values than would be expected if none of the genes were active.

Results from skin cancer experiment

In the skin cancer study (Section 1.2.2), four dual-channel arrays were hybridized, each with a cancerous and normal sample. On the arrays each spot was printed twice. For illustration purposes only, we use all eight values for each gene in each condition as independent replicates.

The Wilcoxon statistic ranges between 36 and 100. 4,608 genes were investigated and their empirical distribution is given by the points in Figure 8.1. The smoothed version of this empirical distribution is given by the dotted line. The heavier tails of the empirical distribution compared to the Wilcoxon distribution (the solid line) give the impression that some of the genes will be differentially expressed.

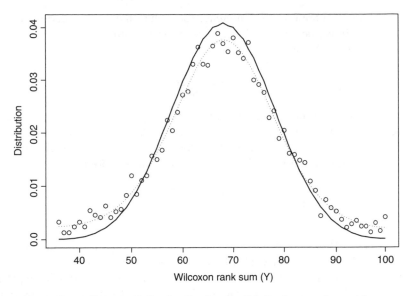

Figure 8.1 The Wilcoxon(8,8) distribution (solid line), together with an empirical distribution for Wilcoxon statistics from a microarray experiment (o) and a smoothed version of this empirical distribution using polynomial Poisson regression (dotted line).

Table 8.3 The number of genes declared 'active' for different values for the posterior distribution.

Posterior cut-off	Number of genes selected
0.900	84
0.950	47
0.990	0
0.999	0

The posterior probability $p(v_g = 0 | z_g > z)$ can be estimated using the Wilcoxon(8,8) distribution for F_0, the empirical distribution for F and 1 for p_0. The result is summarized in Table 8.3.

8.3.2 Bayesian t-test

When only a few replicates are taken, many standard comparison tools, such as traditional t-tests, are not very powerful and can only detect very gross differential expressions. In this section, we focus on alternatives that borrow strength across the genes to make more powerful inference for each of the genes individually. This can be readily implemented via Bayesian hypothesis testing.

Let z_g be the t-statistic,

$$z_g = \frac{\overline{x}_{g1.} - \overline{x}_{g2.}}{se},$$

where se is some estimate of the standard error. If z_g is differentially expressed with fold change δ_g, then z_g is distributed as a translated t-distribution. Assuming that these fold changes δ_g are themselves drawn from a normal distribution with mean 0, we introduce the following approximation of the **likelihood**:

$$z_g | \{v_g = 0\} \sim N(0, \sigma_0^2),$$

$$z_g | \{v_g = 1\} \sim N(0, \sigma_1^2),$$

where $\sigma_0^2 < \sigma_1^2$. Typically, we assume that $\sigma_0^2 = 1$, although for unusual estimates of the standard error se (Efron et al. 2001), this is not necessarily the case. Given the population activation probability p_1, that is, the fraction of differentially expressed genes on the microarray, the distribution of the observed gene expressions can be written as a mixture of normals,

$$z_g \sim (1 - p_1)N(0, \sigma_0^2) + p_1 N(0, \sigma_1^2). \tag{8.14}$$

The model can be represented as a directed acyclic graph (DAG) as in Figure 8.2.

Computational details

The second component of a Bayesian model are the **prior distributions** on the parameters. They represent our 'prior knowledge' about the values of the parameters. Typically, we know very little about these values and therefore we prefer to use non-informative priors. This is achieved in two steps. Firstly, flat priors are used—for instance, the prior on p_1 is taken to be uniform.

$$p_1 \sim \text{Un}(0, 1) \tag{8.15}$$

Notice that given p_1 the v_g's are Bernoulli(p_1) distributed,

$$v_g \sim \text{Bernoulli}(p_1). \tag{8.16}$$

Secondly, hyper-priors β and δ are used in order to avoid making overly strong assumptions on the rate of the inverse gamma distribution for σ_0^2 and σ_1^2 respectively:

$$\sigma_0^{-2} \sim \text{Gamma}(\alpha, \beta) \quad \sigma_1^{-2} \sim \text{Gamma}(\gamma, \delta). \tag{8.17}$$

We note again that, in the case of ordinary t-statistics, σ_0^2 is taken to be the fixed constant $\sigma_0^2 = 1$. The prior distributions for β and δ are taken to be Gamma,

$$\beta \sim \text{Gamma}(g_0, h_0) \quad \delta \sim \text{Gamma}(g_1, h_1). \tag{8.18}$$

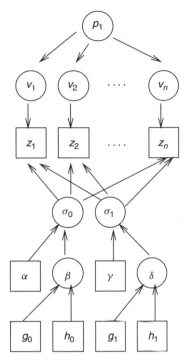

Figure 8.2   A graphical representation of a hierarchical model for differential expression. The graph is a directed acyclic graph (DAG), which makes estimation of the joint distribution particularly simple.

We use conjugate priors for p_1, σ_0^{-2}, σ_1^{-2}, β and δ so that updating them in the Gibbs sampler, which is outlined below, becomes particular easy. The hyper-parameters α and γ are set to 1, just as the hyper-hyper-parameters g_0, h_0, g_1 and h_1.

This completes the specification of the Bayesian model. Given this model, the aim of Bayesian hypothesis testing is to determine the **posterior distribution** of all the parameters,

$$\theta = (\vec{v}, p_1, \sigma_0^2, \sigma_1^2, \text{etc.} \dots),$$

given the data z. Given the careful conjugate structure, a Gibbs sampler is an advantageous choice for estimating the parameters θ. A Gibbs sampler does not calculate the posterior distribution $p(\theta|z)$ explicitly. It simulates draws from this distribution. Using sample summaries, we can get a good sense of the joint posterior distribution as well as of the marginal distribution of interest, $p(v|z)$.

Let θ_i stand for one of the components of the parameter vector $\theta = (\vec{v}, p_1, \sigma_0^2,$ etc.). We denote by θ_{-i} the vector θ from which the component θ_i has been removed,

$$\theta_{-i} = \theta \backslash \{\theta_i\}.$$

The Gibbs sampler samples from the conditional distributions $p(\theta_i|\theta_{-i}\, y)$ until it converges to the stationary distribution. After this burn-in period, each draw $\theta = (\vec{v}, p_1, \sigma_0^2, \sigma_1^2, \text{etc.} \dots)$ is a draw from the posterior $p(\theta|z)$.

Bayes Theorem shows that the conditional distribution $p(\theta_i|\theta_{-i}\, y)$ is proportional to the likelihood times the prior, $p(y|\theta)p(\theta)$, as a function of θ_i. The full likelihood was given by Equation (8.14). The DAG in Figure 8.2 specifies conditional independence relationship, so that the joint prior over all parameters in the model can be written as follows:

$$p(p_1, \vec{v}, \sigma_0^{-2}, \sigma_1^{-2}, \alpha, \beta, \gamma, \delta, g_0, h_0, g_1, h_1)$$

$$= p(p_1)p(\vec{v}|p_1)p(\sigma_0^{-2}|\alpha, \beta)p(\beta|g_0, h_0)p(\sigma_1^{-2}|\gamma, \delta)p(\delta|g_1, h_1).$$

Each of the marginal prior distributions are given by Equations (8.15) to (8.18). It can then easily be deduced that the conditional distributions $p(\theta_i|\theta_{-i})$, needed in the Gibbs sampler, are given as follows:

$$p_1|\theta_{-p_1}, y \sim \text{Beta}\left(1 + \sum v_g, n + 1 - \sum v_g\right)$$

$$v_g|\theta_{-p_1}, y \sim \text{Bernoulli}\left(\frac{\frac{p_1}{\sigma_1}e^{-\left(\frac{z_g}{\sigma_1}\right)^2/2}}{\frac{p_0}{\sigma_0}e^{-\left(\frac{z_g}{\sigma_0}\right)^2/2} + \frac{p_1}{\sigma_1}e^{-\left(\frac{z_g}{\sigma_1}\right)^2/2}}\right)$$

$$\sigma_0^{-2}|\theta_{-\sigma_0^{-2}}, y \sim \text{Gamma}\left(\alpha + \frac{n - \sum v_g}{2}, \beta + \frac{\sum_{v_g \equiv 0} z_g^2}{2}\right)\Big|_{\{\sigma_0^{-2} > \sigma_1^{-2}\}}$$

$$\sigma_1^{-2}|\theta_{-\sigma_1^{-2}}, y \sim \text{Gamma}\left(\gamma + \frac{\sum v_g}{2}, \delta + \frac{\sum_{v_g \equiv 1} z_g^2}{2}\right)\Big|_{\{\sigma_0^{-2} > \sigma_1^{-2}\}}$$

$$\beta|\theta_{-\beta}, y \sim \text{Gamma}\left(\alpha + g_0, h_0 + \sigma_0^{-2}\right)$$

$$\delta|\theta_{-\delta}, y \sim \text{Gamma}\left(\gamma + g_1, h_1 + \sigma_1^{-2}\right)$$

A technical difficulty is the identifiability constraint on the variances, $\sigma_0^2 < \sigma_1^2$. Stephens (2000) has shown that such constraints may slow down mixing of the Gibbs sampler. However, the constraint has the immediate advantage that the output of the sampler is directly interpretable. Moreover, for mixtures with only two components the slow down in convergence is not very serious.

Results from skin cancer experiment

In the skin cancer experiment, Dr Barr compared the expressions of over 4, 600 genes in cancerous tissue versus normal tissue on 4 dual-channel arrays. The mixture model was fitted to the t-statistics of each of the genes. For a full description

Table 8.4   Posterior mean estimates of
the parameters of interest.

Parameter	Cancer data (5,000 burn-in, 6,000 rep)
p_1	0.223
σ_0	0.910
σ_1	2.628

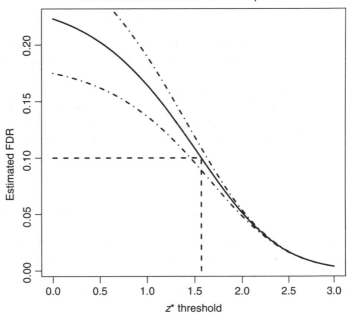

Figure 8.3   The relationship between the estimated false discovery rate and the
selected rejection threshold.

of this study, see Section 1.2.2. It is of interest to find genes that are differentially
expressed between the normal and cancer tissue.

With the posterior mean estimates of p_1, σ_0 and σ_1 in Table 8.4, it is possible
to calculate the posterior 'inactivity' probability,

$$\hat{p}(v_g = 0 \mid z_g \geq z^*) = 0.777 \frac{1 - \Phi(z^*/0.910)}{1 - 0.777\Phi(z^*/0.910) - 0.223\Phi(z^*/2.628)},$$

where Φ is the distribution function of the standard normal distribution. Figure 8.3
shows the posterior probability for different values of z^*. For example, if we declare

Table 8.5 Relationship between the posterior cut-off and the number of genes selected in the skin cancer experiment.

Posterior cut-off	Number of genes selected
0.900	236
0.950	188
0.990	120
0.999	18

all t-statistics with a score higher than 1.57 as 'active', then the estimated false discovery rate is 10%. Approximately 236 genes have a t-statistic higher than that cut-off.

However, the uncertainty about the parameters translates into uncertainty about the estimate. A 95% confidence interval for p_1 is given as $(0.175, 0.272)$; thus the fraction of affected genes on the microarray is expected to lie somewhere between 17.5% and 27.2%. Because of this uncertainty alone, the confidence interval for the FDR lies between 9% and 11% when $z^* = 1.57$, as can be seen in Figure 8.3.

Extensions

It was perhaps not completely satisfactory that the previous model as defined by Figure 8.2 uses a *zero* mean distribution to model the expression levels for genes that are *differentially* expressed. It is more intuitive to model the observed t-statistics z_g as a mixture of three normals: one with a negative mean, the other with a positive mean, while keeping the zero mean distribution reserved for the unexpressed genes. We define a new classification variable:

$$v_g = \begin{cases} 1 & \text{if gene } g \text{ is over-expressed} \\ 0 & \text{if gene } g \text{ is not differentially expressed} \\ -1 & \text{if gene } g \text{ is under-expressed.} \end{cases} \qquad (8.19)$$

The observed gene expressions are defined conditionally on the value of the classification variable:

$$z_g | \{v_g = -1\} \sim N(\mu_{-1}, \sigma_{-1}^2),$$

$$z_g | \{v_g = 0\} \sim N(0, \sigma_0^2),$$

$$z_g | \{v_g = 1\} \sim N(\mu_1, \sigma_1^2),$$

where $\mu_{-1} < 0$ and $\mu_1 > 0$. As mentioned before, $\sigma_0^2 \equiv 1$ in many applications, when z_g is the ordinary t-statistic. If the fraction of under-expressed genes on

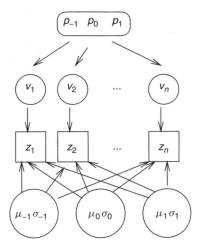

Figure 8.4   A graphical representation of a hierarchical model for differential expression where under- and over-expression are modelled separately.

the microarray of interest is p_{-1}, the fraction of non-expressed gene is p_0, and the fraction of over-expressed is genes p_1, then the **likelihood** of gene expression levels can be written as a mixture of three normals,

$$z_g \sim p_{-1}N(\mu_{-1}, \sigma_{-1}^2) + p_0N(0, \sigma_0^2) + p_1N(\mu_1, \sigma_1^2). \qquad (8.20)$$

Further extensions are possible, such as modelling gene expression as an unknown number of normals with only one fixed component for the inactive genes, but they are computationally much more expensive.

Figure 8.4 shows the DAG for this model, in which under- and over-expression is explicitly modelled by means of separate distributions. The values (p_{-1}, p_0, p_1) represent the fractions of under-, non- and over-expressed genes on the microarray respectively. The parameters v are the classification variables of the individual genes and take values -1, 0 or $+1$.

As before, flat conjugate **priors** are used and hyper-parameters are introduced to avoid making strong prior assumptions. The prior on (p_{-1}, p_0, p_1) is taken to be Dirichlet(1, 1, 1). For the variances, Gamma hyper-priors are taken, for which the shape parameter is taken to be constant and the rate parameter to be Gamma distributed with fixed hyper-hyper-priors. For the means $-\mu_{-1}$ and μ_1, exponential hyper-priors are taken. The mean μ_0 is set to zero. Given the probabilities $p = (p_{-1}, p_0, p_1)$, the v_g's are independently multinomial(1, p) distributed.

A combination of Gibbs steps and Metropolis-Hastings steps will sample from the conditional distributions $p(\theta_i|\theta_{-i}y)$ until it converges to the stationary distribution. Computational details are omitted here. The stationary distribution is the posterior joint distribution of the parameters.

In order to find the rejection region for a fixed FDR of α, we calculate the posterior 'inactivity' probability,

$$\hat{p}(v_g = 0 \mid z_g \geq z) = p_0 \frac{1 - \Phi(z/\sigma_0)}{1 - \sum_{i=-1}^{1} p_i \Phi((z - \mu_i)/\sigma_i)}$$

and reject all genes for which

$$\hat{p}(v_g = 0 | z_g \geq z) \leq \alpha.$$

Palla (2004) discusses this and other Bayesian models for detecting differential expression in more detail.

9

Predicting Outcomes with Gene Expression Profiles

Typically, microarray experiments are performed in order to answer very specific questions. In the skin cancer experiment (Section 1.2.2), one would like to find out which genes vary between skin cancer cell lines and normal cell lines. In the breast cancer experiment (Section 1.2.3), the aim is to establish whether the gene amplification profiles are predictive for the aggressiveness of a breast tumour. And if it is, then one would like to know which genes in particular are responsible for the severity of the cancer. In the mammary gland experiment (Section 1.2.4), the analysis of gene function is of primary interest. That is, one would like to correlate particular stages in the development to the expression of particular genes. For example, at the start of the lactation period, just before the pups begin sucking milk, as well as at the beginning of the involution period, just after the pups stop sucking milk, there is an accumulation of milk in the mammary gland. Finding genes that show marked behaviour during precisely those two periods might give us a clue to which genes are responsible for regulating the lactation.

9.1 Introduction

In many microarray experiments, it is known in advance where the samples come from or what conditions they have been exposed to. In fact, often it is of interest to find out whether the expression profiles can give any insight into the difference between these conditions or classes. For example, Golub et al. (1999) considered patients' expression profiles to distinguish between different types of leukemia.

Under *classification*, we understand the study of discriminating observations $x_1, x_2, \ldots, x_n$ into pre-specified classes $C(x_1), C(x_2), \ldots, C(x_n)$. There are a host of methods that can achieve this objective. Golub et al. (1999) proposed a weighted

Statistics for Microarrays: Design, Analysis and Inference   E. Wit and J. McClure
© 2004 John Wiley & Sons, Ltd   ISBN: 0-470-84993-2

voting scheme, with similarities to a form of discriminant analysis (Dudoit et al. 2000). Dudoit and Fridlyand (2003) give a detailed comparison of several classification methods, such as nearest neighbour classification, linear and quadratic discriminant analysis, classification trees and aggregating classifiers. Their general conclusion is that several simple methods have a robust performance. This is supported by theoretical considerations in Section 9.3.1. Therefore, the focus of this chapter will be on simple methods, such as (diagonal) linear discriminant analysis (LDA) and k-nearest neighbour classification with a robust choice for k.

9.1.1   Probabilistic classification theory

Most of the classification methods can be described as special implementations of a general probabilistic classification analysis, known as *Bayes classifiers*. The decision rule for classifying x_i as one of the k classes, $C_1, \ldots, C_k$, depends on three things:

- prior information about the class frequencies $\pi_1, \ldots, \pi_k$,

- information about how class membership affects the gene expression profiles, x_i $(i = 1, \ldots, n)$,

- misclassification costs $L(i, j)$ of classifying an observation x which belongs to C_i into class C_j.

Our aim is to use a classification rule $\mathcal{R}$, $\mathcal{R}(x) \in C = (C_1, \ldots, C_k)$, that minimizes the expected misclassification costs,

$$\mathcal{R} = \arg \min E_{\mathcal{R}} L$$

$$= \arg \min \sum_{i=1}^{k} \sum_{j=1}^{k} L(i, j) \ p(\mathcal{R}(x) = C_j \mid C(x) = C_i) \ \pi_i. \tag{9.1}$$

As can be seen from Equation (9.1), the expected misclassification costs depend on the misclassification cost function and the prior probabilities of class memberships. We shall show that the probability $p(\mathcal{R}(x) = C_j | C(x) = C_i)$ for a certain class of cost functions depends intrinsically on the likelihood of the data given the class memberships and the prior class probabilities.

Costs of misclassification: misclassification rate

In principle, it is possible to specify a fully general misclassification cost function, $L(i, j)$, of classifying a sample from class i into class j. For example, it might be more serious to misclassify an expression profile from an aggressive breast cancer tumour than it is to misclassify one from a benign tumour.

A typical class of cost functions that are considered in practice is the class of functions for which the misclassification costs only depend on the type of class that is misclassified, that is,

$$L(i, j) = \begin{cases} 0 & i = j \\ L_i & i \neq j. \end{cases} \qquad (9.2)$$

A special case of this cost function is one that sets each L_i to some constant, that is,

$$L_i = 1. \qquad (9.3)$$

The expected misclassification costs under the cost function in (9.3) is known as the *expected misclassification rate*. Ripley (1996) showed that the classification rule that minimizes the expected misclassification costs of the type (9.3) is the one that assigns each sample to the class that has the highest posterior probability.

Theorem 9.1 *The classification rule that minimizes the expected misclassification cost for cost functions specified by (9.3) is given by*

$$R(x) = C^* \Leftrightarrow C^* = \arg\max_C p(C \mid x).$$

In other words,

$$R(x) = C^* \Leftrightarrow C^* = \arg\max_C p(x|C)\pi_C. \qquad (9.4)$$

This rule is known as Bayes rule.

In the unusual case where the prior class probabilities π and the likelihood $p(x|C)$ are known, an optimal classifier is available. The classifier, given by (9.4), is optimal in the sense that it minimizes the expected misclassification rate.

Most classifiers used in practice can be regarded as ways to estimate the posterior directly, such as nearest neighbour classification, or as a way to estimate the likelihood, such as in the case of discriminant analysis.

Estimation of the misclassification rate is an important aspect of building a classifier. If a training set is used to fine-tune the parameters of an estimator, then the performance of that estimator on that set cannot be trusted to give a good estimate of a predictive misclassification rate.

There are two main ideas in estimating the misclassification rate. The most honest, if slightly wasteful, procedure is to split the data into a training and validation set. The classifier is built using only the training set and then evaluated on the evaluation set to predict the misclassification rate. A problem with this procedure is that on the one hand one would like to use a large training set to make the classifier as good as possible, whereas on the other hand the validation set should also be large in order to get a good estimate of the misclassification rate.

A solution to this problem has been suggested in the form of *K-fold cross-validation*. This procedure partitions the data into K equal-sized, randomly selected groups of samples. The predictor is iteratively trained on $K - 1$ of the groups, and the misclassification rate is calculated on the remaining one. The final estimate of the misclassification rate is estimated by the average of the K misclassification values. If K is set to n, the number of samples in the study, this procedure is known as *leave-one-out cross-validation*. The final predictor in the case of K-fold cross-validation is based on the full data set.

Values of K in K-fold cross-validation close or equal to n have been shown to lead to biased (typically, anti-conservative) estimates of the misclassification rate. Breiman (1996) has suggested 10-fold cross-validation as a robust choice, although others believe that 4- or 5-fold cross-validation is a sensible choice for the small data sets typically encountered in microarray studies.

Prior information about class membership

Information about the relative frequencies of the class memberships may some-times be available, particularly in epidemiological and certain clinical settings. For example, a particular tumour may be more common than others. Certain muscular diseases are more common than, for example, the Gillain–Barré syndrome. It is therefore sensible to accept the verdict that a particular expression profile represents Gillain–Barré only if there is a lot of evidence in its favour.

In most instances, however, it is quite unfeasible to say anything about the prior probabilities. Unusual sampling schemes might result in relative class fre-quencies that are very different from the population prior probabilities. Hospitals that are known for a special treatment for a certain type of cancer, might expect to see relatively many more patients with that type of cancer than the population frequencies would suggest.

The likelihood: the conditional probability of the data

The *training set* is the collection of samples for which both the gene expressions and the class memberships are known. These samples are used to train the classifier. This can mean estimating the posterior probability of class membership directly or estimating the probability of the data given the class memberships, known as the *likelihood*. This quantity appears in Bayes rule (9.4), and therefore is essential for minimizing the expected misclassification rate.

The most commonly used likelihood in practice is derived from the normal distribution. In many applications, it is reasonable to assume that the observations are some form of aggregate and therefore subject to the central limit theorem. Additive aggregate values tend to be normally distributed, whereas multiplicative aggregates tend towards the log-normal distribution. Gene expressions are positive quantities where fold changes, that is, multiplicative aggregation, are the rule. Artifacts apart, it is therefore not unreasonable to expect that the log-transformed expression quantities are normally distributed (see Sections 6.2.1 and 6.2.2 for an extensive discussion on this issue).

For the likelihood of the class memberships, the probability of the gene expressions should be considered collectively. If each of the genes follows a univariate normal distribution, then the collection of all the genes possesses a multivariate normal distribution. Whereas each of the univariate distributions is characterized by its mean and variance, a multivariate normal distribution has the additional complication of a covariance structure between the genes. For large numbers of genes, the variance–covariance matrix becomes too large to be estimated with any reliability and further simplifying assumptions such as independence of gene profiles are needed.

Predictor evaluation

Evaluating the behaviour of a predictor, such as its misclassification rate, can be a tricky business. Clearly, the misclassification rate on the training set can give a very distorted view of the true misclassification rate. A very complex classifier might perfectly classify all training samples but is very unlikely to generalize to other samples.

Honest misclassification rates can be obtained by applying the predictor to a validation set that is not used in training the data. This approach risks setting a lot of data aside for validation, which cannot be used for training the predictor. Cross-validation methods are designed to deal with this.

Predictor p-value.   In some cases where there is sufficient data relative to the number of classes and predictors, it is possible to set aside a validation set. This validation set can be used to evaluate the behaviour of the predictor, in particular, whether it predicts the sample class memberships better than random.

Let L and V be the *learning* (or training) set and *validation* set respectively. There are K possible classes. The misclassification rate of a predictor $C(., L)$ can be estimated on the basis of the validation set as the fraction of misclassified samples,

$$\widehat{\text{MCR}}\,(C(., L)) = \frac{1}{|V|} \sum_{x \in V} 1_{\{C(x,L) \neq C(x)\}},$$

where $C(x)$ is the true class membership of sample x. This fraction can be regarded as the test statistic for testing whether the predictor $C(., L)$ behaves better than random.

$H_0 :\ C(., L)$ behaves in a 'random' fashion

$H_1 :\ C(., L)$ behaves differently from 'random'.

We assume that a 'random' predictor would randomly assign the validation samples to the classes according to the relative frequencies with which the classes occur in the training set. Therefore, under the null hypothesis, the estimated misclassification rate is approximately the ratio of a binomial distribution and V,

$$(\widehat{MCR} \times V)|H_0 \sim Bi(|V|, \theta),$$

where θ is the misclassification probability for a single validation sample $x \in V$, that is,

$$\theta = P(C(x, L) \neq C(x))$$

$$= \sum_{k=1}^{K} P(C(x, L) \neq k) P(C(x) = k)$$

$$= \sum_{k=1}^{K} (1 - \pi_k^L) \pi_k^V.$$

Here, π_k^L and π_k^V are the relative class frequencies in the learning and validation set respectively. A p-value for the statistical evidence against the null hypothesis, when a misclassification rate of q is estimated on the basis of the validation set, can therefore be obtained via

$$p\text{-value} = P_{H_0}(\widehat{MCR}(C(., L)) \leq q)$$

$$= \sum_{i=0}^{q|V|} \binom{|V|}{i} \theta^i (1 - \theta)^{|V|-i}$$

$$\approx \Phi \left(\frac{|V|(q - \theta) + 0.5}{\sqrt{|V|\theta(1 - \theta)}} \right),$$

where Φ is the cumulative distribution function of a standard normal distribution. As in traditional hypothesis testing, p-values less than a certain cut-off α, typically $\alpha = 0.05$, provide evidence in favour of the predictor and against the null hypothesis H_0.

K-fold cross-validation.   Setting aside a validation set is typically only possible when there is a lot of data available. In the microarray context, this is often not the case. However, if the whole data set is used for training the predictor, it might not seem obvious how an honest misclassification rate can be obtained. An elegant procedure is to omit part of the data in turn, fit the model to each subset and obtain predictions for each omitted observation. This is known as K-fold cross-validation.

The method begins by fitting the predictor to all available data to obtain the predictor $C(., L)$. Building the predictor might involve a certain operational procedure, such as, for instance, stepwise regression or determining the value of some tuning parameter λ. This procedure is described by a set of rules, such as 'omit a gene from the feature set if its associated p-value is larger than 0.05.' We indicate these rules, irrespective of their actual implementation in building $C(., L)$, as

PROC. Cross-validation is a method for estimating the misclassification rate of a predictor built according to the procedure *PROC*.

First, it randomly partitions the training set into K equally or almost equally sized subsets, $L_1, L_2, \ldots, L_K$. Then it builds K predictors $C(., L \backslash L_k)$ according to the same procedure *PROC* but on the basis of the learning set L from which the kth subset L_k has been omitted. Then it estimates the misclassification rate of the procedure *PROC* that built $C(., L)$, as

$$\mathrm{MCR}(PROC) = \frac{1}{|L|} \sum_{k=1}^{K} \sum_{x \in L_k} 1_{\{C(x, L \backslash L_k) \neq C(x)\}},$$

where $C(x)$ is the true class of the observation x. Clearly, the predictors $C(., L \backslash L_k)$ might be very different from $C(., L)$. They might not even use the same genes. However, they do arise from using the same procedure. Therefore, theoretically speaking, the misclassification rate estimate is associated with the procedure, rather than with the actual predictor.

The choice of K is a matter of debate. A common choice is $K = |L|$. This is also known as *leave-one-out* cross-validation. A problem of *leave-one-out* cross-validation is its computational burden. Except in special circumstances, it requires building $|L|$ predictors $C(., L \backslash \{x_k\})$. Another problem is that the $|L|$ training sets $L \backslash \{x_k\}$ are very similar and the cross-validation estimate of the misclassification rate has a high variance. Twofold cross-validation is the other extreme. In that case, the cross-validation predictors are built on only half of the available data and might not be very good. The cross-validation estimate of the misclassification rate therefore tends to be overestimated, that is, biased. In general, Hastie et al. (2001) recommend using five- or tenfold cross-validation as good compromise between bias and variance of the cross-validation estimate of the misclassification rate.

9.1.2 Modelling and predicting continuous variables

Classification can be regarded as predicting the class labels of the observations. Sometimes the quantity of interest is not the class labels but some continuous response related to the observations. For example, for each of the tumours in the breast cancer example a severity score, known as Nottingham prognostic index (NPI) score, is recorded. It is of considerable interest to find out if there is a genetic component to the NPI, and if so, which genes are informative of the NPI.

Modelling and predicting a continuous variable is the realm of regression models, survival analysis and related methods. This field has been under development ever since Legendre and Gauss almost simultaneously invented least squares in the early nineteenth century. Recently, the field was given a new impulse by the introduction of penalized methods. These methods are particularly suited to deal with large number of variables. This means that these methods can be very useful in a microarray context.

Modelling of continuous variables can be broken up into several distinct but interrelated steps. First, there is the issue of deciding on a *potential set of features*, that is, the subset of genes (or meta-genes; *cf.* Section 9.2) that might be used in the modelling process. Typically, one would like to consider no more variables than there are observations, although this is not necessary in some of the penalized methods. Then, if a certain set of variables is selected, one goes on to *model fitting*. Traditionally, least squares or (penalized) likelihood have been used for this purpose. For linear models explicit formulas exist, whereas for most non-linear models iterative methods or approximations are required. Finally, there is the issue of *model validation*. In this step, goodness-of-fit and predictive ability of the model are considered. This might involve the issue of *variable selection*, that is, deciding on the subset of features that are actually used in the modelling process. Akaike's information criterion (Akaike 1973) or the Bayesian information criterion are methods for deciding the number of variables or the value of any other tuning parameter. This parameter could also involve *model selection* more generally, such as, for example, the nature of the functional relationship between the features and the response.

9.2   Curse of Dimensionality: Gene Filtering

High-density microarrays confront the analyst with an embarrassment of riches. There are so many possible classifiers, that, if all genes were used to classify a small number of samples in distinct classes, then this would typically always be possible by chance, whether or not the data are informative at all. This is the so-called *curse of dimensionality*. For example, in a two-dimensional space it is always possible to separate two observations perfectly; the same thing is true for three or fewer points in a three-dimensional space and for n or fewer points in an n-dimensional space. This observation has been forcefully made in Amaratunga and Cabrera (2004) and elsewhere.

9.2.1   Use only significantly expressed genes

There have been several suggestions to remedy this problem. Dudoit and Fridlyand (2003) restrict their attention to a fixed number of genes that have the highest ratio of between- to within-groups sums of squares. This corresponds to taking the genes with the smallest p-value in an ordinary one-way ANOVA setting,

$$x_i = \alpha_{C(x_i)} + \epsilon_i,$$

where $C(x_i)$ is the class label of the gene profile of sample i. For the case of only two classes, this approach is equivalent to finding the genes with the small p-value in an ordinary two-sample t-test. Nguyen and Rocke (2002b), Radmacher et al. (2002), describe similar gene filtering approaches.

Most authors select a pragmatic number of genes so as to make the number of variables smaller than the number of samples, whereas Amaratunga and Cabrera

(2004) fix a certain significance level and reduce the number of genes further by means of selecting a number of principal components, which is again smaller than the number of samples.

The main attraction and purpose of these methods is that it results in a traditional classification problem, where the number of variables is less than the number of observations.

There are a number of problems associated with this approach. First of all, there is the issue of bias. For example, in a completely random gene expression data set with thousands of genes and a relatively small number of samples, this gene selection procedure will find 'statistically significant' genes. These genes, thus selected, are expected to do a good job in classifying the data set. Nevertheless, the classification does not have any predictive power.

The only way in which this procedure is valid is if the genes are selected only on the basis of the training set rather than on the full data set and if a validation set is used to estimate the misclassification rate. If the 'significant genes' were selected on the basis of the whole data set and subsequently reused for estimating the misclassification rate, a far too rosy picture of the procedure would be painted.

Example 9.1 We simulate a completely random gene expression matrix for 50 samples and 5,000 genes by repeatedly drawing values from a standard normal distribution. The samples are randomly split into two classes, completely unrelated to their 'expression values'.

Then, we select 10 genes with the highest ratio of the between- versus within-group sums of squares *on the basis of the whole data*. A training set is created by randomly taking 25 out of 50 samples. For this training set, a predictor is built using the gene in this set of 10 genes that performs best in a logistic regression. The predictor is then applied to the validation set of the remaining 25 samples. The misclassification rate is a misleading small 0.16.

The misclassification rate would have been a much more representative 0.48, if the 10 genes were instead selected on the basis of the highest ratio of the between-versus the within-group sums of squares *within the training set*.

Another problem with using only significantly expressed genes is that the method selects only a particular type of predictor. By selecting genes that have small within-sums-of-squares compared to between-sums-of-squares, it selects, by necessity, predictors that are essentially univariate and 'linear'.

Bo and Jonassen (2002) suggest considering pairs of predictors. For binary classification problems, pairs of genes can be selected via Hotelling's multivariate t-test (Hotelling's T^2 test). This test considers whether or not the gene expressions $(x_{g1i}, x_{g2i})_{i=1,\ldots,n_x}$ of genes g_1 and g_2 under the first condition can be considered to have the same mean as the expressions $(y_{g1i}, y_{g2i})_{i=1,\ldots,n_y}$ of the same genes under the other condition. The test statistic of Hotelling's T^2 test is defined as

$$T^2 = ((\bar{x}_{g_1}, \bar{x}_{g_2}) - (\bar{y}_{g_1}, \bar{y}_{g_2}))S^{-1}((\bar{x}_{g_1}, \bar{x}_{g_2}) - (\bar{y}_{g_1}, \bar{y}_{g_2})),$$

where S is the estimate of the variance–covariance matrix. Conditional on their being no difference between the two means, T is distributed (Hotelling 1931) as,

$$T^2|H_0 \sim 2F_{2;n_x+n_y-2},$$

where n_x and n_y are the respective sample sizes. If it is a multinomial classification problem, one can use a generalization of the Hotelling's T^2 test, such as MANOVA, to find pairs of the most-informative genes.

Considering pairs of genes rather than individual genes provides a richer set of potential predictors that might involve gene–gene interactions and inhibitors. However, it is always important to realize that pre-selection of features of this type always tends to select genes that have a unimodal behaviour over the classes of interest. We do not contend, in the case where there are few observations, that this is a sensible thing to do. One of the few classification methods that is able to deal with multimodal features is nearest neighbour classification. However, feature selection is also a problem for this method when the number of genes is much larger than the number of samples.

9.2.2 PCA and gene clustering

Another recurring suggestion in classification and prediction problems is to reduce the feature space by using *meta-genes*. These are representative genes taken from the whole data set via a variety of methods. We consider approaches that use principal component analysis (PCA) and related methods, as well as gene clustering as a way to find potential predictors.

Spang et al. (2002) introduce the concepts of *super-genes*, defined as the eigen-genes that result from a singular value decomposition of the gene expression matrix (*cf.* 7.1.3). A very similar approach is suggested in Amaratunga and Cabrera (2004). They consider the principal components of the feature space, both before and after they have reduced the space by means of selecting only differentially expressed genes.

Unsurprisingly, Amaratunga and Cabrera (2004) find that the first few principal components of the whole data are uninformative, whereas the first principal components of the significant features cluster the data almost perfectly. The reason is that in the first case, there is no guarantee that the main 'directions' in the gene profiles correspond to the particular classification of interest—unless the classes differ massively across the board from one another. When the genes are *first* reduced to the set of genes that differ significantly across the conditions and *then* used for classification, the main directions clearly do correlate with the classification—by design. In fact, the same difficulty in estimating the misclassification rate arises as before: the genes are selected on the basis of all the data, and it is therefore not surprising that they will be good at classifying those data. In order to estimate the misclassification rate appropriately, a subset of the data should be set aside before one selects the significant genes to reduce the feature set and build the classifier.

Another way of reducing the feature space is by clustering genes and deriving the *meta-genes* by finding a representative gene in each cluster. Unlike in the

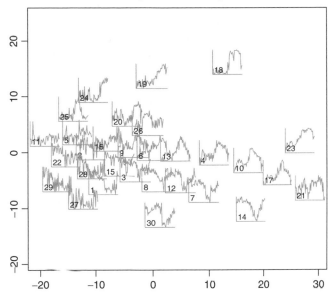

Figure 9.1 Partitioning-around-medoids clustering with a fixed number of groups ($k = 30$) applied to the mammary gland data set as a method to reduce the dimensionality of the prediction problem.

case of PCA, there is no guarantee that the first meta-genes coming from gene clustering would explain most of the variation of the feature set. On the other hand, these meta-genes have the advantage that they are directly associable with a group of genes, whereas the PCA super-genes are sometimes incomprehensible linear combinations of genes.

In practice, meta-genes can be found with any centroid-based clustering algorithm. We suggest using PAMSAM (*cf.* Section 7.3.2), which is a robust cousin of the k-means algorithm, with a fixed value k for the number of gene clusters. It is reasonable to use for k a value in the order of a third to a half of the number of samples, n, available for classification or prediction. The reduced feature set of $k < n$ cluster medoids can then be used in any traditional classification algorithm. These algorithms can involve further feature selection steps. This is no problem as long as some form of validation of the final answer takes place, either by means of a separate validation set or via K-fold cross-validation.

In the mammary gland data set, there are 54 arrays available. A feature set with 30 dimensions is on the edge of feasibility if we also would like to use a validation set to estimate honest misclassification rates. Figure 9.1 shows the result of a k-medoids clustering via PAMSAM with $k = 30$. The profiles shown can be used as the potential feature set in any classification or prediction problem.

9.2.3 Penalized methods

One of the crucial problems in fitting a model with more variables than observations is that the design matrix is singular. Solutions that implicitly or explicitly depend on the calculation of the inverse of the design matrix are therefore doomed from the very start. It was noticed that by tinkering with the design matrix, effectively by adding penalty terms to it, the problem became identifiable again, and it was possible to find a unique solution. This method has been successfully applied to smoothing problems and to traditional linear regression problems.

Hastie et al. (2001) introduce a penalized form of discriminant analysis. No explicit gene filtering is needed as the algorithm guarantees 'identifiability' of the solutions. Penalized discriminant analysis (PDA) penalizes the mean and variance parameters for the expression of each gene. The PDA coefficients are idealized high-dimensional decision boundaries. A problem with PDA is that even if the resulting predictor has a good performance, it might be hard to pinpoint which genes were responsible for this.

For regression problems, Tibshirani (1996) proposed a *least absolute shrinkage and selection operator* (LASSO). This is a penalized linear regression set-up. Assume that we have a continuous response of interest y and profiles $x_1, \ldots, x_p$ for n samples over p genes, where the number of predictors p is large and the number of observations n is small. The usual aim is to fit

$$y_i = \alpha + \sum_{j=1}^{p} \beta_j x_{ij} + \epsilon_{ij},$$

in some kind of 'optimal way', typically by least squares. Tibshirani (1996) twists this by adding in the additional constraint that the sum of the absolute values of the parameter estimates cannot exceed a certain value C,

$$\sum_{j=1}^{p} |\beta_j| \leq C.$$

We discuss this method in more detail in Section 9.4.

9.2.4 Biological selection

Finally, it is possible that the biologist has other reasons to believe that the true features are only a subset of the number of genes on the slide. If there are good biological reasons to limit the number of predictors, then this is a welcome relief to the computational burden of the bioinformatician.

An example of biological selection might include considerations of significant alteration across the experimental conditions. For example, if a particular gene has not shown any significant changes in expression in the whole experiment, then it might not be of biological interest at any rate. What constitutes significance might change from problem to problem. Figure 9.2 shows the gene expression

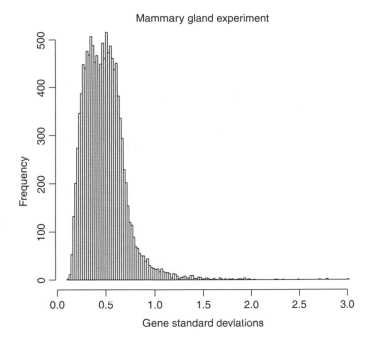

Figure 9.2   Histogram of standard deviations of gene expressions for each gene across the 54 arrays in the mammary gland experiment. Genes with small standard deviations can be left out from the feature set in prediction problems, if such is deemed sensible.

standard deviations of all genes in the mammary gland experiment across all 54 microarrays. There is a suggestion of a bimodal distribution. The first component might correspond to merely technical variation, whereas the second component might exhibit biological variation. An idea might be to leave out all genes with a standard deviation less than 0.4. This corresponds to almost 5,000 genes and ESTs in the data set. Other options could be to use the Absence/Presence calls in Affymetrix software. In general, these methods should be used only sparingly as they are rather blunt tools.

9.3   Predicting Class Memberships

9.3.1   Variance-bias trade-off in prediction

In practice, a predictor is built on the basis of a training set of data. Because the training set is, at best, a random sample from the study population, the training set and therefore the constructed predictor are variable. For a particular learning set L, the predicted class for a fixed gene profile x based on the predictor $P(L)$ is the random variable $C(x, L)$. If there are K possible classes, then

$$C(x, L) \in \{1, \ldots, K\}.$$

If the membership likelihoods and population proportions of the classes were known, then the optimal predictor is a Bayes Rule, B (cf. Section 9.1.1). For a fixed gene profile x, the predicted class based on Bayes is $C_B(x) \in \{1, \ldots, K\}$. Notice that the predicted class is not variable. In particular, it does not depend on the training set. Given that the predicted class is variable, it make sense to consider the bias and variance of the predicted posterior probabilities $\widehat{P}(C(x) = k|x; L)$.

It is a common issue in estimation that there is a direct trade-off between the variance and the bias of an estimator. Reducing the bias is a desirable property, but this can sometimes be accompanied by a large increase in the variance of the estimator. Making estimators more complex in order to reduce the bias is therefore not always a worthwhile thing to do if one is interested in controlling the expected squared error,

$$E(\hat{\theta} - \theta)^2 = E(\hat{\theta} - E\hat{\theta})^2 + E(E\hat{\theta} - \theta)^2$$

$$= \text{Variance} + \text{Bias}^2.$$

This issue is exacerbated for prediction. Friedman (1996) decomposes the misclassification rate for a simple binary classification problem into a contribution by the misspecification of the predictor and by the random nature of the future observation, that is, Bayes risk,

$$P(C(x, L) \neq C(x)) = |2P(C(x)|X = x) - 1| \, P(C(x, L) \neq C_B(x))$$

$$+ P(C_B(x) \neq C(x)).$$

Friedman shows that the classification error depends in a more complicated way on the variance term of the predictor of the posterior class membership probability.

For example, in a particular binary prediction problem, two classes have membership densities that are normally distributed with a common variance of 1 and a mean of 0 and 2 respectively. Both classes are equally common. The true posterior probability for an observation x is given as

$$P(C(x) = 1|x) = \frac{p(x|C(x) = 1)\pi_1}{p(x|C(x) = 1)\pi_1 + p(x|C(x) = 0)\pi_0}$$

$$= \frac{e^{-\frac{1}{2}(x-2)^2}}{e^{-\frac{1}{2}(x-2)^2} + e^{-\frac{1}{2}x^2}}.$$

Training the predictor on a variable training set, L, will result in an estimate of $p(C(x) = 1|x)$ indicated by $P(x, L)$. This quantity is variable and has a mean and variance,

$$E[P(x, L)] = \mu_x$$

$$V[P(x, L)] = \sigma_x^2.$$

The impact of the bias $P(C(x) = 1|x) - \mu_x$ and the variance σ_x^2 can be derived under some special assumptions. Friedman shows that if the predictor is normal,

$$P(x, L) \sim N(\mu_x, \sigma_x^2),$$

then the misspecification part of misclassification rate has the following form,

$$|2P(C(x)|X = x) - 1|\Phi\left(\text{sign}\left(P(C(x)|X = x) - \frac{1}{2}\right)\frac{1/2 - \mu_x}{\sigma_x}\right),$$

where Φ is the c.d.f. of the standard normal distribution. Assuming that the predictor is unbiased, that is, $P(C(x) = 1|X = x) = \mu_x$, then this misspecification contribution to the misclassification rate reduces to

$$\text{'Misspecification rate'} = |2\mu_x - 1|\Phi\left(\text{sign}\left(\mu_x - \frac{1}{2}\right)\frac{1/2 - \mu_x}{\sigma_x}\right).$$

Figure 9.3 shows the relationship of the misspecification rate versus the classification variance and the posterior probability of the classification $P(C(x) = 1|x)$.

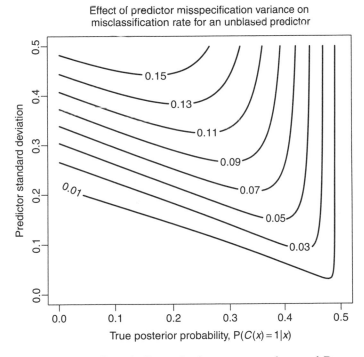

Effect of predictor misspecification variance on misclassification rate for an unbiased predictor

Figure 9.3   The contour lines indicate the increase over the usual Bayes risk of a predictor if the predictor is unbiased but has some variance. The increase in the misclassification rate is shown for different values of the true posterior probability of classifying an observation x in class 1.

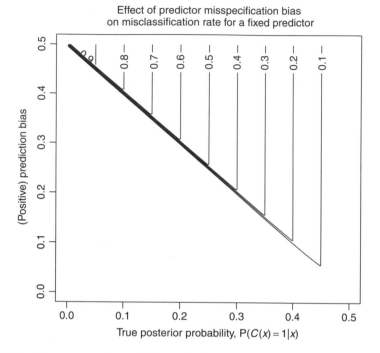

Figure 9.4   The contour lines indicate the increase over the usual Bayes risk of a predictor if the predictor does not vary under different learning sets L but does however have a certain bias, $\mu_x - P(C(x) = 1|x)$. The increase in the misclassification rate is shown for different values of the true posterior probability of classifying an observation x in class 1.

Because of equal population proportions for both classes, the plot is symmetric in $P(C(x) = 1|x) = 0.5$. It shows how typically for a fixed predictor variance, the misspecification rate increases. However, for values close to the decision boundary $P(C(x) = 1|x) = 0.5$, the decision is a coin flip anyway and is not influenced by the misspecification variance.

On the other hand, if the predictor has some bias but no variance, then the resulting 'misspecification rate' is shown in Figure 9.4. It shows that for a large range of values of the bias,

$$0 \leq \text{bias} < 0.5 - P(C(x) = 1|x),$$

the contribution of the bias to any misclassification is null. This is good news for simple classifiers that may have some bias but little variance: even though the classifier consistently under- or overestimates the posterior membership probabilities, it can still result in an excellent classifier.

9.3.2 Linear discriminant analysis

Linear discriminant analysis (LDA) has been found to be an excellent classifier in many complex classification situations. In the STATLOG project, which compared and evaluated a range of classification techniques, LDA was found to be among the best classifiers for many of the data sets considered (Michie et al. 1994). Dudoit and Fridlyand (2003) found that in microarray situations LDA also has a very robust performance.

There are alternative discriminant analysis techniques, such as quadratic discriminant analysis. However, (Michie et al. 1994, p.192) recommends the use of linear discriminants, unless the covariances between the different classes are very different and the sample size is very large; note that this latter requirement is unlikely to be met in microarray studies.

Mathematical details

The main assumption of discriminant analysis is that the class densities can be written as multivariate Gaussians. This means that if $x = (x_1, \ldots, x_p)$ is the gene profile for a sample from class k, then the probability density is given as

$$p(x|C = k) = \frac{1}{(2\pi)^{p/2}|\Sigma_k|^{1/2}} e^{-\frac{1}{2}(x-\mu_k)^t \Sigma_k^{-1}(x-\mu_k)},$$

where μ_k is the mean expression vector and Σ_k is the covariance matrix for class k. *Linear* discriminant analysis arises when it is assumed that all classes have the same covariance matrix Σ. This assumption is particularly useful if there are few training samples. By combining the estimates across the classes, a more robust estimate of the covariance matrix can be achieved.

If the number of features or genes p is in the same order of the number of training samples, then even estimation of a common covariance matrix with p^2 elements is out of reach. In such cases, it is only feasible to estimate a common diagonal covariance matrix. This corresponds to *diagonal* LDA and implies making an additional assumption of independence of genes within each class.

LDA can be seen as a direct application of Bayes rule. If the aim is to minimize the misclassification rate, then Bayes rule assigns the sample x to the class k that maximizes the posterior probability,

$$p(C = k|x) = \frac{p(x|C = k)\pi_k}{\sum_{i=1}^{K} p(x|C = i)\pi_i}.$$

By explicitly defining the class densities, LDA aims to maximize

$$\log p(C = k|x) = \log \frac{\frac{\pi_k}{(2\pi)^{p/2}|\Sigma|^{1/2}} e^{-\frac{1}{2}(x-\mu_k)^t \Sigma^{-1}(x-\mu_k)}}{\sum_{i=1}^{K} \frac{\pi_i}{(2\pi)^{p/2}|\Sigma|^{1/2}} e^{-\frac{1}{2}(x-\mu_i)^t \Sigma^{-1}(x-\mu_i)}}$$

$$= C + \log \pi_k - \frac{1}{2}(x - \mu_k)^t \Sigma^{-1}(x - \mu_k),$$

where C is some fixed, and therefore irrelevant, constant.

In practice, the class density parameters are unknown and need to be estimated from the training set. Let x_i be the gene profile of the ith sample in the training set belonging to class $C(i)$. Let n_k be the number of samples in class k in the training set, let there be K classes and let n be the size of the training set. Then, for LDA, we can estimate the parameters via

$$\hat{\pi}_k = n_k/n,$$

$$\hat{\mu}_k = \frac{1}{n_k} \sum_{C(i)=k} x_i,$$

$$\hat{\Sigma} = \frac{1}{n-K} \sum_{k=1}^{K} \sum_{C(i)=k} (x_i - \hat{\mu}_k)(x_i - \hat{\mu}_k)^t$$

(Hastie et al. 2001). In the case of *Diagonal* LDA, the estimates for the relative class frequencies π_k and class means μ_k remain the same. However, the matrix Σ is a diagonal matrix with entries $\sigma_1^2, \ldots, \sigma_p^2$, whose components can be estimated via the usual sample variance,

$$\hat{\sigma}_j^2 = \frac{1}{n-K} \sum_{k=1}^{K} \sum_{C(i)=k} (x_{ij} - \hat{\mu}_{kj})^2.$$

It is unlikely that the assumptions of normality are satisfied in practice. However, the effect of model-misspecification on class membership prediction is small in practice. Nevertheless, the estimation of the covariance matrix Σ can be affected quite significantly by the presence of outliers. Therefore, 'robustifying' estimates of the covariance matrix can be useful.

Finding genes: stepwise discriminant analysis

If the feature set is known, then LDA is a straightforward method to classify samples, as discussed above. However, in most microarray settings, the feature set of interest is not known. In fact, it is often a major issue to determine which genes are responsible for the class memberships of the samples. For example, in the breast cancer experiment, it is of interest to determine which genes might be involved in the severity of the breast tumour. One way to determine the severity is by determining retrospectively which patients died as a result of the breast cancer. This is a binary classification problem with an unknown feature set.

Stepwise LDA is a method that combines a gene selection procedure with ordinary discriminant analysis. It is initialized by finding the gene with the lowest leave-one-out cross-validation misclassification rate in the training set. It then iteratively adds the gene to the feature set that reduces the overall cross-validation error the most. If no reduction in the leave-one-out cross-validation score is found, then the procedure stops. The classifier is then defined by the LDA on the selected gene set.

As the leave-one-out cross-validation misclassification rate is used to build the optimal predictor, it is not a good guide for the actual misclassification rate. It might be too optimistic if the potential feature space is too large. Therefore, in order to evaluate the misclassification rate of the classifier, it is important to perform an overall cross-validation of the whole procedure or to apply the classifier to an independent validation set.

Stepwise discriminant analysis in R: step.lda.   Stepwise discriminant analysis is implemented in R in the step.lda function. It requires a training set train together with the class membership vector class. Optionally, one can include a test-set test for which class memberships are predicted. A true validation set can be set aside by specifying the fraction of validation samples in valid.frac. It is also possible to obtain a misclassification rate via K-fold cross-validation.

Example: mammary gland experiment

In the mammary gland experiment, the main focus of attention is to study the crucial genes that trigger the transitions to the main developmental events in the mammary gland of mice. When a female mouse develops from a pup into an adult after some 12 weeks, it is ready for the next main developmental event, namely pregnancy. During pregnancy, the ducts in the mammary gland begin to proliferate. The ducts further thicken in the lactation phase, whereas after lactation, in the so-called involution of the mammary gland, the network of ducts begins to thin out, reverting back to the adult state before pregnancy.

In this experiment, there are 12,488 probe sets representing approximately 8,600 genes. These probe sets are measured over 54 arrays. Our first aim is to reduce the potential feature set to a workable set of data. Almost 5,000 ESTs and genes are eliminated since they do not alter substantially across the 54 arrays (standard deviation less than 0.4; *cf.* Figure 9.2). The remaining 7,802 genes and ESTs are further reduced via clustering the genes into 30 groups via a k-medoids algorithm (*cf.* Figure 9.1).

Then, the set of 54 observations is split into a set of 36 training samples and a set of 18 validation samples. The aim is to find the genes that separate efficiently the virgin samples, the pregnancy samples, the lactation samples and the involution samples. We perform gene selection and classification with stepwise LDA. Gene clusters 1, 10, 18 and 25, each represented by their medoid, are found to yield the best separation between the developmental stages. Figure 9.5 shows the four cluster medoids that are most informative.

In order to evaluate the classifier, we apply it to the 18 validation samples. This results in an 11% misclassification rate. The expected misclassification rate for a random classifier would be

$$\left(1 - \frac{8}{36}\right) \frac{4}{18} + \left(1 - \frac{14}{36}\right) \frac{7}{18} + \left(1 - \frac{6}{36}\right) \frac{3}{18} + \left(1 - \frac{8}{36}\right) \frac{4}{18} = 0.72,$$

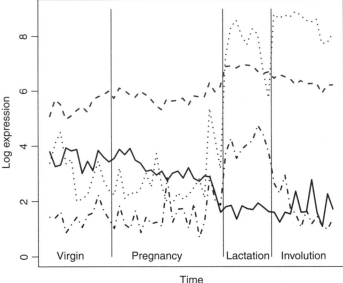

Figure 9.5   Stepwise LDA in the mammary gland experiment selects four genes to separate the four developmental stages on the basis of minimizing the cross-validation error. The profiles of these four genes can clearly be seen to separate the four developmental stages.

that is, 72%. The null hypothesis that the selected classifier is no better than random can be rejected with a lot of confidence (p-value = 10^{-7}). The resulting classification of the training samples and the validation samples is shown in Figure 9.6. The plot shows a two-dimensional Sammon representation of the space spanned by the four selected genes. It shows clearly how the genes separate the involution and lactation phase from the other two. The fact that the virgin phase and pregnancy phase do not seem clearly separated is the result of a slightly misleading two-dimensional representation; one is in fact behind the other.

Example: breast tumour experiment

In the breast tumour experiment, it is of interest to determine which genes influence the severity of breast cancer. The DNA amplification patterns of 59 genes is studied in 62 tumours. Follow-up information on the patients revealed that 20 patients died of breast cancer, whereas 35 did not. For 7 patients, no follow-up information was available. It is of interest to discover which variables might be a good predictor for separating those patients that die as a result of breast cancer and those that do not.

The data for which the cause of death is available is split randomly into a training set (44 tumours) and a validation set (11 tumours). The 7 tumours for

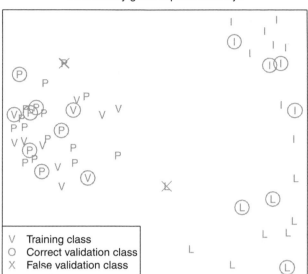

Figure 9.6   Stepwise LDA misclassifies 2 out of the 18 (11%) validation samples. The plot is a Sammon representation of the four dimensional feature space.

which no information is available are taken as the test set for which a prediction is generated.

Stepwise LDA selects two genes (Nos. 1 and 23). This classifier only misclassifies 1 of the 11 tumours (9%) in the validation set. The resulting prediction is given in Figure 9.7. For the particular make-up of the training and validation set, a 'random' classifier based on the prior probabilities in the training set would have achieved a 44.8% misclassification rate. With a p-value of 0.014, there is a lot of evidence to believe that the predictor is better than random.

9.3.3 k-nearest neighbour classification

k-nearest neighbour classifiers act on the assumption that samples with almost the same feature values should probably belong to the same class. More precisely, given a set of genes $g_1, \ldots, g_m$ known to be important for the membership of a particular class, a k-nearest neighbour classifier assigns an unclassified sample to the class that is most prevalent among the k samples whose expression values for these m genes are closest to those in the sample of interest.

Typically, the gene expression profile $x_i = (x_{i1}, \ldots, x_{im})$ for sample i is compared with other profiles using Euclidean distance (Hastie et al. 2001),

$$d(x_i, x_j) = \left\{ \sum_{k=1}^{m} (x_{ik} - x_{jk})^2 \right\}^{1/2}.$$

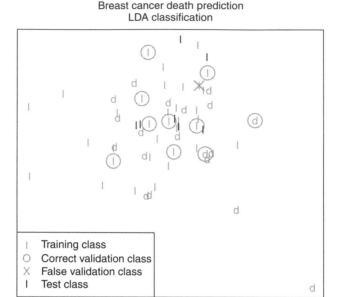

Breast cancer death prediction
LDA classification

	Training class
O	Correct validation class
X	False validation class
I	Test class

d = Death from breast cancer,   I = Alive or death from other causes

Figure 9.7   Stepwise LDA misclassifies 1 out of the 11 (9%) validation samples. The plot is a Sammon representation of the two selected genes.

However, it is possible to use any of the distance measures that are discussed in Section 7.1.2. In particular, the 'correlation' distance, $1 - \rho$, and the Manhattan distance can be useful.

Instead of modelling the likelihood $p(x|C(x) = i)$ as discriminant analysis does, the aim of k-nearest neighbour classification is to estimate the posterior probability $P(C(X) = i|X = x)$ of a gene profile x belonging to class i directly. For a particular choice of k, it estimates this probability as the relative fraction of samples that belong to class i among the k samples with the most similar expression profiles,

$$\hat{P}(C(X) = i|X = x) = \frac{1}{k} \sum_{j=1}^{k} 1_{\{C(x_{(j)}) = i\}},$$

where $x_{(j)}$ is defined as the jth nearest observation in the learning set L to x,

$$d(x_{(1)}, x) \leq d(x_{(2)}, x) \leq \ldots \leq d(x_{(n)}, x).$$

Despite its simple appearance, the decision boundaries of a k-nearest neighbour classifier can be highly irregular. It is an essentially non-linear classifier that can accommodate multiple regions for the same class. In fact, 1-nearest neighbour classification is so irregular that it creates a little region for every single observation in the learning set. Such irregular classifiers have typically a poor prediction performance. Larger values for k result in more regular decision boundaries and classifiers with a smaller variance. The number of neighbours k is a tuning parameter and is best determined via cross-validation.

Finding genes: stepwise k-nearest neighbour

If the feature set is known and the tuning parameter k has been determined, then k-nearest neighbour classification is straightforward. Unfortunately, this is typically not the case in microarray studies as the number of potential predictors is very large and selection of variables is an integral part of the scientific problem.

Buttrey (1998) and Buttrey and Karo (2002) propose a stepwise approach to select variables within a k-nearest neighbour classification setting. They propose a sequential forward (or backward) selection algorithm, which adds (or deletes) a variable based on leave-one-out cross-validation. The selection process stops when adding (or deleting) any variable does not improve the cross-validated misclassification rate.

A word of caution: as the leave-one-out cross-validation misclassification rate is used to build the optimal predictor, it is not a good guide for the actual misclassification rate. In order to obtain an honest estimate of the misclassification rate, K-fold cross-validation on the whole stepwise procedure or the application of the classifier to an independent validation set is required.

Stepwise k-nearest neighbour in R: step.knn.   Within the smida library, stepwise k-nearest neighbour classification is implemented in the step.knn function. It requires a training set train together with the class membership vector class. Optionally, one can include a test set test for which class memberships are predicted. A true validation set can be set aside by specifying the fraction of validation samples in valid.frac. It is also possible to obtain a misclassification rate via K-fold cross-validation.

The knnTree library by Buttrey and Karo (2002) contains the function knn.var, which performs stepwise k-nearest neighbour classification. It is part of a larger function to perform k-nearest neighbour classification within each leaf of a classification tree.

Example: breast tumour experiment

The breast tumour experiment, described in Section 1.2.3, contains the DNA amplification patterns for 62 breast cancer tumours over 59 genes. The main point of the study is to determine which genes influence the severity of breast cancer. Follow-up information on the patients revealed that 20 patients died of breast cancer, whereas 35 did not. For 7 patients no follow-up information was available. We use stepwise k-nearest neighbours to determine which variables are predictive of breast cancer death and apply the predictor to the 7 unclassified observations.

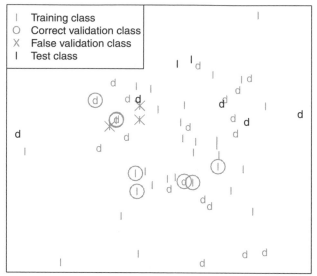

Breast cancer death prediction
K–nearest neighbours classification

d = Death from breast cancer,   I = Alive or death from other causes

Figure 9.8   Sammon plot of the 5 predictive genes for the 44 training samples, 11 validation samples and 7 test samples; 3 of 11 validation samples are misclassified (27%).

The selected predictor, shown in Figure 9.8, uses $k = 3$ nearest neighbours for prediction. Stepwise k-NN selects five genes (Nos. 44, 46, 47, 55 and 59). The selected variables are different from the two that stepwise LDA selected (Section 9.3.2). This k-NN classifier misclassifies 3 of the 11 tumours (27%) in the validation set, as opposed to only 1 misclassification by stepwise LDA. The resulting prediction is given in Figure 9.8.

The training set consists of 14 and 30 tumours associated with people that did and did not die of breast cancer respectively. In the validation set, 6 people died of breast cancer and 5 did not. A random classifier would have misclassified a random observation in the validation set with probability

$$(1 - 14/44)5/11 + (1 - 30/44)6/11 = 0.483.$$

Therefore, the p-value for the hypothesis that the selected predictor is random is approximately

$$p\text{-value} = P(M \leq 3) = 0.136,$$

where the number of misclassifications M under the null hypothesis that the predictor is 'random' has an approximate binomial(11, 0.483) distribution. We cannot, therefore, reject the null hypothesis with the usual significance level $\alpha = 0.05$.

9.4   Predicting Continuous Responses

A host of methods for relating explanatory variables to noisy continuous responses has emerged over the course of the past century, ranging from linear methods such as multiple regression to non-linear methods such as loess regression and other smoothing techniques.

The trade-off between rigid linear methods and flexible non-linear methods is very similar to the trade-off between simple and complex techniques for classification. Linear methods may not capture the full dynamic range of the actual effect; in terms of prediction, however, linear methods with a high bias but low variance can easily outperform more complex competitors with low bias but high variance.

Here, we shall discuss two methods. The first is a penalized regression method that is particularly suitable to deal with a large number of predictors. The second is a simple non-linear method called k-nearest neighbour regression.

Many other methods do exist for continuous prediction and have been applied to microarray data. For example, partial least squares (PLS) is a linear regression method whereby one looks for directions in the space of the explanatory variables that have high variance *and* a high correlation with the response. Nguyen and Rocke (2002a) describe the application of PLS in a proportional hazards model to partially censored survival data.

9.4.1   Penalized regression: LASSO

Let the vector $y = (y_1, \ldots, y_n)$ be the responses associated with a feature set $\mathcal{X}$. In typical microarray applications, the number of parameters p is of the same order as, or larger than, the number of observations n. If the independent variables $x_1, \ldots, x_p$ are linear predictors of y, then we can write the response as a linear combination of $x_1, \ldots, x_p$,

$$y_j = \alpha + \sum_{i=1}^{p} \beta_i x_{ij} + \epsilon_j.$$

It is perfectly possible that one or more 'predictors' are not predictive at all. In that case, their associated coefficients are zero.

A common way to estimate the coefficients is via a method called *least squares*, which selects those values for β that minimize the total sum of squared deviations from the model to the observations. This simple method is extremely powerful when there is a sufficient number of observations. However, when the number of observations is in the same order as the number of predictors, then this estimation procedure can be very unstable.

Consider, as an example, the data in Table 9.1. The response was generated from x_1 with noise,

$$y_i = x_{1i} + \epsilon_i, \quad \epsilon_i \sim N(0, 0.15^2), \quad i = 1, \ldots, 4,$$

Table 9.1   Simulated data set with one response y and two explanatory variables x_1 and x_2. The response y is generated as a noisy version of x_1.

y	1.25	1.99	3.14	4.12
x_1	1.23	1.82	2.94	4.23
x_2	0.97	0.99	1.03	1.04

Table 9.2   Model fit of $y = \beta_1 x_1 + \epsilon$ to the example data.

	Estimate	Standard. Error	t value	P($>$|t|)
$\hat{\beta}_1$	1.0145	0.0288	35.22	0.0001

Table 9.3   Model fit of $y = \beta_0 + \beta_1 x_1 + \epsilon$ to the example data.

	Estimate	Standard Error	t value	P($>$|t|)
$\hat{\beta}_0$	0.2045	0.1933	1.06	0.4009
$\hat{\beta}_1$	0.9479	0.0690	13.73	0.0053

where $x_i \sim N(i, 0.2^2)$. The data x_2 are completely unrelated to the prediction problem. In fact, $x_2 \sim N(1, 0.03)$ independent of x_1 and y.

Table 9.2 shows the effect of fitting a simple model without intercept and x_2. The true model, $E(y|x_1) = x_1$, is within the margins of error of the estimates. However, when an intercept and then x_2 are added to the linear model, the estimates begin to fluctuate wildly, as can be seen in Table 9.3 and Table 9.4.

Table 9.4   Model fit of $y = \beta_0 + \beta_1 x_1 + \beta_2 x_2 + \epsilon$ to the example data.

	Estimate	Std. Error	t value	P($>$|t|)
$\hat{\beta}_0$	−11.9490	4.4429	−2.69	0.2266
$\hat{\beta}_1$	0.6405	0.1172	5.46	0.1153
$\hat{\beta}_2$	12.8426	4.6938	2.74	0.2231

The reason for such large fluctuations can be understood via a geometric argument. When the number of variables and the number of observations are almost the same, then the surface that one can fit to the observations is too flexible. In fact, by tilting the plane dramatically in the x_2 direction, a marginally better fit can be obtained—which is, however, not statistically significant anymore.

The main idea behind penalized regression is to introduce a penalty term for fluctuations that are too large. *Ridge regression*, for example, is defined as the usual least squares solution with an additional constraint, that is,

$$\hat{\beta}^{\text{ridge}} = \arg\min_{\beta} \sum_{i=1}^{n} (y_i - \beta_0 - \sum_{j=1}^{p} \beta_i x_{ij})^2,$$

subject to

$$\sum_{j=1}^{p} \beta_i^2 \leq c,$$

for some fixed tuning parameter $c > 0$. Ridge regression *penalizes* coefficients that are too large. It is a shrinkage method that, depending on the size of c, shrinks the coefficients to zero.

LASSO regression, introduced by Tibshirani (1996), shares many similarities with ridge regression. It is a form of multiple linear regression that aims to minimize the sums of squares of the errors, but subject to a slightly different constraint,

$$\hat{\beta}^{\text{lasso}} = \arg\min_{\beta} \sum_{i=1}^{n} (y_i - \beta_0 - \sum_{j=1}^{p} \beta_i x_{ij})^2,$$

where

$$\sum_{j=1}^{p} |\beta_i| \leq c, \tag{9.5}$$

for some fixed tuning parameter $c > 0$. The difference between Ridge regression and the LASSO can best be understood by considering Figure 9.9. Minimizing the sum of squared errors *on* the Ridge constraint, that is, the circle, leads to a solution that is typically somewhere in the plane $\mathcal{R}^2$. The LASSO constraint, on the other hand, is a diamond centred around the origin. Minimizing the sum of squared errors *on* the LASSO constraint, therefore, leads to a solution, where typically one of the coefficients is put to zero. A way to interpret this property of the LASSO estimates is as a means of variable selection.

Figure 9.10 shows the parameter estimates for the example data in Table 9.1 for different choices of the constraint c in Equation (9.5), ranging from $c = 0$ to the completely unconstrained solution for which the values for β correspond to the ordinary least squares estimate.

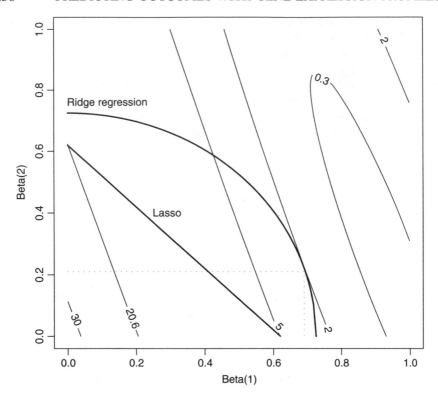

Figure 9.9   The LASSO constraint $|\beta_1| + |\beta_2| \leq 0.62$ and the ridge constraint $\beta_1^2 + \beta_2^2 \leq 0.73$ are shown together with several contour lines of the sums of squared errors for the example data.

Cross-validation for determining c

The value of c in Equation 9.5 can be determined by K-fold cross-validation. This means that the data set L is partitioned into K random, almost equal-sized subsets, $L_1, \ldots, L_K$. For a fixed value $c > 0$ and for each observation $x \in L_k$ ($k \in \{1, \ldots, K\}$), a prediction

$$\hat{y}_c(x) = \hat{\beta}_0^{(-k)} + \sum_{i=1}^{p} \hat{\beta}_i^{(-k)} x_i,$$

is proposed, where $\hat{\beta}^{(-k)}$ is the LASSO estimate of β based only on $L \backslash L_k$ under the constraint $\sum_{i=1}^{p} |\beta_i^{(-k)}| \leq c$. The quantity

$$SS(c) = \sum_{x \in L} (y(x) - \hat{y}_c(x))^2 \tag{9.6}$$

is the sum of squares of the cross-validated prediction errors. By minimizing (9.6) over c, a value for the tuning parameter can be found.

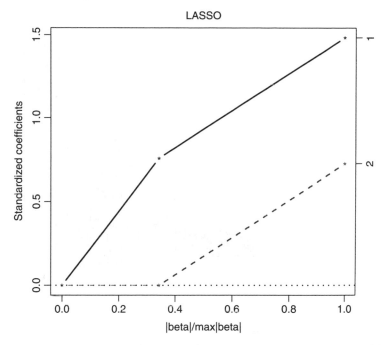

Figure 9.10 The relationship between the LASSO estimates of the coefficients, β_1 and β_2, on the y-axis and the relative size of the constraint on the x-axis.

Leave-one-out cross-validation is a special case of K-fold cross-validation. It is typically very computationally intensive as it involves n ($n = |L|$) applications of the learning method. However, Hastie et al. (2001) shows that for linear procedures, for which the fit is a linear transformation of the response, that is,

$$\hat{y} = Hy,$$

generalized cross-validation provides a convenient approximation to leave-one-out cross-validation. The generalized cross-validation error rate is a one-step approximation of the leave-one-out cross-validated error rate in Equation 9.6,

$$GCV(c) = \sum_{x \in L} \left(\frac{y(x) - \hat{y}_c(x)}{1 - \text{trace}(H)/n} \right)^2,$$

where $\hat{y}_c(x)$ is the estimated response using the full LASSO with tuning parameter c.

Figure 9.11 shows the generalized cross-validation error rate for the example data set in Table 9.1. A minimum lies somewhere between relative values of 0.1 and 0.3 for the constraint. From Figure 9.10, it is clear that this corresponds to a model for which the second coefficient is zero, which corresponds to the actual data generating procedure.

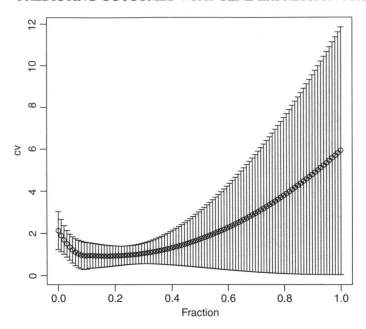

Figure 9.11   The generalized cross-validated sums of squared errors for the example data set. A minimum is found somewhere between 0.1 and 0.3.

Although generalized cross-validation is much less computationally burdensome, its error estimates will still have the large variance that leave-one-out cross-validation estimates have. Using K-fold cross-validation with $K = 5$ or $K = 10$ should be considered when important conclusions are at stake.

Breast cancer example: determinants of survival

In the 1990s, a combination score, named the Nottingham Prognostic Index (NPI), was created in order to assess the severity of a breast tumour (Galea et al. 1992). In the breast tumour data set, both the survival time and the NPI score were known for 45 of the 62 patients. The survival time has been measured as the length of time from the biopsy until a breast cancer related death. For 25 out of the 45 patients, the survival time is censored, for example, owing to the fact that the patient is still alive or has died of other causes. Figure 9.12 shows the relationship between the survival time and the NPI score. A rough pattern is that the larger the NPI score, the lower the survival time.

It was of interest to the researcher of the study to find out whether DNA amplification of certain genes could be a better predictor for survival than the NPI score. Figure 9.13 shows the LASSO coefficient estimates for increasing values of the penalty term. The choice of the penalty threshold is determined via generalized cross-validation. Figure 9.14 shows the cross-validation error for different choices

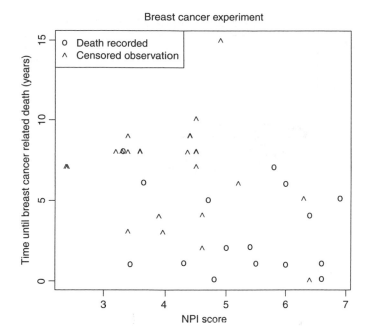

Figure 9.12   A negative relationship between NPI score and survival time.

of the threshold. Values between 0 and 0.23 seem reasonable choices for c in Equation (9.5). We choose $c = 0.05$ as a cut-off.

There are some interesting variables among those whose coefficient is not equal to zero at $c = 0.05$. They are indicated by the black lines and variable numbers in Figure 9.13. For example, variable 60 is tumour size—the only non-genetic variable. Its parameter is negative, which suggests that the larger the tumour the shorter the survival time. Gene 23 is an interesting variable, as it also seemed to be predictive of breast cancer death *per se* in Section 9.3.2.

Implementation of LASSO in R

The R library lars written by B. Efron and T. Hastie contains the functions lars and cv.lars. Both functions require a matrix x of potential predictors and a continuous response vector y. The default regression type is the "lasso", but it is also possible to perform two other types of penalized regression, namely least angle regression ("lar") and forward stagewise additive modelling ("forward. stagewise"). See Efron et al. (2004) and Hastie et al. (2001) for details of these two methods. The lars objects that result from these three regression types can be visualized via the plot.lars function. In particular, the parameter estimates for different choices of the tuning parameter can be shown.

In order to decide a proper cut-off for the tuning parameter, the cv.lars function does a generalized cross-validation for a grid of values of this parameter.

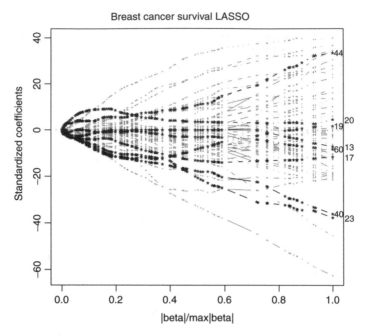

Figure 9.13 LASSO estimates for the 59 gene amplification profiles and tumour size (variable 60) in predicting survival time after biopsy. The seven selected genes (and tumour size) are numbered in the right margin of the plot.

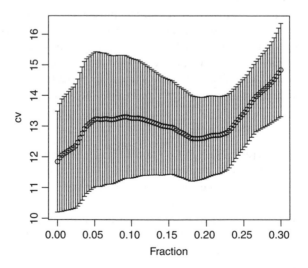

Figure 9.14 Generalized cross-validation for the tuning parameter in the LASSO breast cancer survival regression. Choices between 0 and 0.23 seem reasonable candidates.

It generates a plot, such as in Figure 9.14, for which tuning parameter with the smallest cross-validation error can be estimated.

9.4.2   k-nearest neighbour regression

The LASSO is a linear method, closely related to multiple linear regression. Sometimes, such linear methods are not appropriate, and it is necessary to consider more flexible models. Many non-linear methods are, however, extremely complex. This model complexity is typically an undesirable feature from the point of view of prediction.

A non-linear method that produces consistently good results is k-nearest neighbours. We encountered k-nearest neighbour classification in Section 9.3.3. In the continuous case, the method predicts the response as the mean of the k-nearest neighbours.

Let d be any dissimilarity measure. **k-nearest neighbour regression** predicts the response y for a statistical unit with covariate x as

$$\hat{y}(x) = \frac{1}{k} \sum_{j=1}^{k} y(x_{(j)}),$$

where $x_{(j)}$ is defined as the jth nearest observation in the learning set L to x, that is,

$$d(x_{(1)}, x) \leq d(x_{(2)}, x) \leq \ldots \leq d(x_{(n)}, x).$$

The method is essentially the same as the k-nearest neighbour imputation algorithm discussed in Section 4.2.3. This algorithm imputes missing values among the covariates by replacing them with the average value for that gene among the k-nearest neighbours.

Stepwise k-nearest neighbour regression

When the number of potential predictors is large, that is, $p \gg 40$ (Hastie et al. 2001), many variable subset selection procedures become infeasible. However, *forward stepwise selection* is still a possibility. It can be used in combination with k-nearest neighbour regression.

The procedure starts out by selecting the variable together with a value of k that has the lowest leave-one-out cross-validated sum of squared errors. It then adds new predictors sequentially, continuously adjusting k, as long as it reduces the cross-validated sums of squares.

For testing the overall validity of the selected k-nearest neighbour model, we can use K-fold cross-validation or test the model on an independent validation set V. Either way, we aim to test the null hypothesis,

H_0: the selected K-NN regression model is not predictive.

As a test statistic, we take the sum of squared errors when we apply the selected
k-nearest neighbour regression model to the validation set. We obtain,

$$SS(\text{model}) = \sum_{x \in V} (y(x) - \hat{y}(x, L))^2.$$

If the model is not predictive, then in fact any choice of k values from the response
of the training set should yield more or less the same sum of squared errors as
$SS(\text{model})$. By bootstrapping a large number of such sums of squared errors,
one can obtain a bootstrap p-value for the above null hypothesis. The p-value is
the empirical quantile at which $SS(\text{model})$ lies among the bootstrapped sums of
squares.

Stepwise k-nearest neighbour in R: `step.knn`.   The function `step.knn` that
performs stepwise k-nearest neighbour classification is also able to deal with k-
nearest neighbour regression. It requires a training set `train` together with the
continuous response vector `class`. Optionally, one can include a test set `test` for
which responses are predicted. A validation set can be set aside by specifying the
fraction `valid.frac` of validation samples to be used for this purpose. It is also
possible to obtain a misclassification rate via K-fold cross-validation.

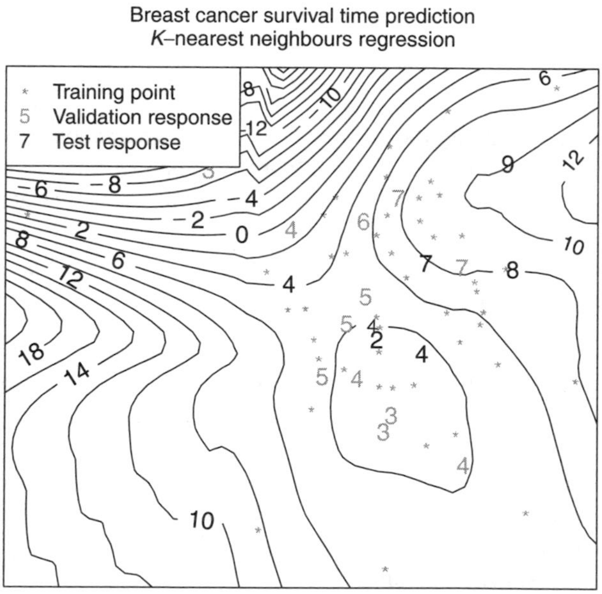

Breast cancer survival time prediction
K–nearest neighbours regression

Figure 9.15   Stepwise k-nearest neighbour regression on the survival time (in
years) in the breast cancer experiment. Three genes (Nos. 13, 20 and 23) are
selected a predictive. The displayed survival times have been rounded to the nearest
year.

Breast cancer experiment: determinants of survival

K-nearest neighbour regression was applied to the breast cancer data set. It was of interest to find out which genes are predictive of the survival time of the patients. In total, there are 62 patients and the survival time after biopsy is known for 58 of them. The stepwise k-nearest neighbour algorithm selected $k = 3$ as well as three genes for prediction. All of these predictors were also selected by LASSO.

A Sammon mapping of the three-dimensional feature space with superimposed contour lines for the survival time are shown in Figure 9.15. When testing the overall validity of the model on a subset of 12 tumours that were set aside for validation, a p-value of 0.014 was found. This means that the k-nearest neighbour regression model is highly significant.

10

Microarray Myths: Inference

Just as in microarray data handling, several 'myths' about microarray inference have gained currency. The aim of this chapter is to discuss some of these myths and, in some cases, expose the absurdities contained in them. The chapter is organized as a loose collection of sections, each dealing with a different microarray inference myth.

10.1 Differential Expression

Although hypothesis testing is widely used, there are certain aspects about its interpretation that are often misunderstood. This section deals with some of these aspects.

10.1.1 Myth: 'Bonferroni is too conservative'

When testing multiple hypotheses, a number of different procedures exist to control several different error rates. Bonferroni is a very well-known procedure. However, it is often disparaged for being 'too conservative'. When testing n hypotheses, the Bonferroni procedure rejects, for a given familywise error rate (FWER) α, each hypothesis whose associated p-value is less than α/n. It is often felt that this is too stringent, and Bonferroni is discarded as a 'poor' error control procedure.

However, this is to slight a procedure that controls perfectly the error rate it is supposed to control—the FWER. The FWER is the probability of falsely rejecting one or more hypotheses when testing multiple hypotheses.

Whilst it is true that certain other FWER control methods have greater *power* than the Bonferroni procedure, there will be little difference between the methods in most microarray settings. For example, in Hochberg's procedure (see Section 8.2.3) $\alpha/(n - g + 1)$ will be little different from α/n when n is large and g—the number

Statistics for Microarrays: Design, Analysis and Inference   E. Wit and J. McClure
© 2004 John Wiley & Sons, Ltd   ISBN: 0-470-84993-2

of hypotheses rejected by Hochberg's procedure—is relatively small. For example, if $n = 10\,000$ and g is $2\,500$, then Bonferroni with $\alpha = 0.05$ would give 5.0×10^{-6} as a decision boundary, whereas Hochberg's procedure would use 6.7×10^{-6} as a cut-off.

What people often tend to mean when criticizing Bonferroni is that they do not feel that the FWER is the appropriate error rate to control. If that is the case, then they should choose to control another type of error. Either the false discovery rate (FDR) or the false positive rate (FPR) may be more suitable.

10.1.2   FPR and collective multiple testing

Although the false positive rate (FPR) is used standardly when testing single hypotheses, somehow it is frowned upon when testing multiple hypotheses. The intuition is that when testing many hypotheses simultaneously while controlling for the FPR, you cannot be certain that *all* of the rejected null hypotheses are in fact false.

This prejudice against the FPR in multiple testing is not necessary as long as one realizes that not all rejected hypotheses have to be false. For example, if testing 100 hypotheses, controlling the FPR at 0.05, 14 hypotheses might be rejected. Even if all the 100 null hypotheses were true, we would expect only about 5 rejections. Therefore, with the additional information that on average at most 5 hypotheses are falsely rejected, the 14 rejected hypotheses are a significant contribution to scientific knowledge.

10.1.3   Misinterpreting FDR

One of the error rates that is useful when testing multiple hypotheses is the false discovery rate (FDR). The FDR is defined as the expected proportion of falsely rejected hypotheses F_P out of the total rejected S,

$$FDR = E(F_P/S).$$

In the context of differential expression, it *roughly* stands for the average number of false discoveries among all genes declared 'active'.

If, for example, 40 genes are rejected using an FDR of 20%, it is tempting to conclude that 'the expected number of false discoveries in the 40 genes selected is 8'. This is incorrect. Care must be taken not to interpret the FDR as being conditional on the actual number of rejections found. The FDR is not defined as a conditional expectation, that is,

$$FDR \neq E(F_P/S|S = s).$$

Despite its appeal, this interpretation is wrong. It is quite possible that for a particular number of rejections, the conditional expectation is quite different from the unconditional expectation. The FDR can be seen as a weighted average of the conditional expectations,

$$FDR = \sum_{s=1}^{\infty} E(F_P/S|S = s)p(S = s).$$

The real interpretation of the FDR is a bit more abstract:

The false discovery rate associated with a decision rule is the average fraction of false discoveries whenever that decision rule is repeatedly applied to a set of hypotheses.

The FDR stands for the average fraction of false rejection associated with a *decision rule*, rather than with any particular number of rejected hypotheses. When fixing a FDR decision rule, such as the Benjamini & Hochberg FDR procedure, the number of rejected hypothesis is still regarded as a random variable.

10.2 Prediction and Learning

10.2.1 Cross-validation

In many cases, the main objective in classifying microarray samples is to minimize the misclassification rate of a classifier. In order to obtain the classifier, it needs to be trained on a set of data, typically called the training set L. Training means estimating the parameters in the classifier so that afterwards for any class c and for any gene profile x, one can calculate the posterior probability of belonging to class c, that is, $P(C(X, L) = c|X = x)$.

The danger of determining the misclassification rate on the basis of the training set is that the model might overfit the data and so an overly rosy picture of the model's performance can be painted. For this reason, one can use a validation set or cross-validation to estimate the misclassification rate (*cf.* Section 9.1.1).

It would be tempting to use cross-validation for several different classifiers and choose the one that minimizes the cross-validated misclassification rate. In itself this is not a bad idea, as long as one remembers that the misclassification rate thus found is not an accurate estimate of the true misclassification. Especially, if this is done over a large class of predictors, the minimum cross-validated misclassification rate tends to underestimate the misclassification rate. In order to achieve an honest evaluation of the classification rate, one could, for example, apply the classifier with the lowest cross-validated misclassification rate to a validation set.

Bibliography

Affymetrix 2001 *Affymetrix Microarray Suite 5.0; User's Guide*. Affymetrix Inc., Santa Clara, CA.

Akaike H 1973 Information theory and an extension of the maximum likelihood principle. In *Proceedings of the Second International Symposium on Information Theory* (eds. Petrov BN and Caski F). Akademiai Kiado, Budapest, pp. 267–81.

Alter O, Brown PO and Botstein D 2000 Singular value decomposition for genome-wide expression data processing and modelling. *PNAS* **97**(18), 10101–6.

Amaratunga D and Cabrera J 2001 Analysis of data from viral DNA microchips. *Journal of the American Statistical Association* **96**(456), 1161–70.

Amaratunga D and Cabrera H 2004 *Exploration and Analysis of DNA Microarray and Protein Array Data*. John Wiley & Sons, New York.

Angulo J and Serra J 2003 Automatic analysis of DNA microarray images using mathematical morphology. *Bioinformatics* **19**(5), 553–62.

Bagchi S and Cheng CS 1993 Some optimal designs of block size two. *Journal of Statistical Planning and Inference* **37**, 245–53.

Bakewell D and Wit EC 2004 Weighted analysis of microarray gene expression using maximum likelihood. Technical Report 04-5, Department of Statistics, University of Glasgow, UK.

Banfield JD and Raftery AE 1993 Model-based Gaussian and non-Gaussian clustering. *Biometrics* **49**, 803–21.

Bassett Jr DE, Eisen MB and Boguski MS 1999 Gene expression informatics—it's all in your mine. *Nature Genetics Supplement* **21**, 51–5.

Benjamini Y and Hochberg Y 1995 Controlling the false discovery rate: a practical and powerful approach to multiple testing. *Journal of the Royal Statistical Society Series B* **57**, 289–300.

Bo TH and Jonassen I 2002 New feature subset selection procedures for classification of expression profiles. *Genome Biology* **3**(4), research:0017.1–0017.11.

Bolstad BM, Irizarry RA, Astrand M and Speed TP 2003 A comparison of normalization methods for high density oligonucleotide array data based on bias and variance. *Bioinformatics* **19**(2), 185–93.

Box G, Jenkins GM and Reinsel G 1994 *Time Series Analysis: Forecasting and Control*, 3rd edn. Prentice Hall, Upper Saddle River, NJ.

Breiman L 1996 Bagging predictors. *Machine Learning* **24**, 123–40.

Bretz F, Landgrebe J and Brunner E 2003 Efficient design and analysis of two colour factorial microarray experiments. Technical Report 2003-1, DNA Microarray Facility, Faculty of Medicine, University of Göttingen, Göttingen, Germany, http://www.cmis.csiro.au/IAP/Spot/spotmanual.htm?

Bryan J, Pollard KS and van der Laan MJ 2002 Paired and unpaired comparison and clustering with gene expression data. *Statistica Sinica* **12**, 87–110.

Buckley MJ 2000 The Spot user's guide. Technical Report, CSIRO Mathematical and Information Sciences. http://www.cmis.csiro.au/IAP/Spot/spotmanual.htm.

Buhler J, Ideker T and Haynor D 2000 Dapple: improved techniques for finding spots on DNA microarrays. Technical Report, UWCSE Technical Report UWTR 2000-08-05, Department of Computer Science and Engineering, University of Washington, Seattle, WA, USA.

Buttrey S 1998 Nearest-neighbor classification with categorical variables. *Computational Statistics & Data Analysis* **28**, 157–69.

Buttrey SE and Karo C 2002 Using k-nearest neighbor classification in the leaves of a tree. *Computational Statistics & Data Analysis* **40**(1), 27–37.

Casella G and Berger RL 2001 Hypothesis testing in statistics. *International Encyclopedia of the Social and Behavioral Sciences*, Vol. 10. Pergamon Press, New York, pp. 7118–21.

Chen Y, Dougherty ER and Bittner ML 1997 Ratio-based decisions and the quantitative analysis of cDNA microarray images. *Journal of Biomedical Optics* **2**(4), 364–74.

Chipman H, Hastie TJ and Tibshirani R 2003 Clustering microarray data. In *Statistical Analysis of Gene Expression Microarray Data* (ed. Speed T), Interdisciplinary Statistics. Chapman & Hall/CRC, Boca Raton, FL, pp. 159–200.

Chollet P, Amat S, Belembaogo E, Curé H, de Latour M, Dauplat J, Bouedec GL, Mouret-Reynier MA, Ferrière JP and Penault-Llorca F 2003 Is Nottingham prognostic index useful after induction chemotherapy in operable breast cancer? *British Journal of Cancer* **89**, 1185–91.

Chu S, DeRisi J, Eisen M, Mulholland J, Botstein D, Brown PO and Herskowitz I 1998 The transcriptional program of sporulation in budding yeast. *Science* **282**(5389), 699–705.

Churchill GA 2002 Fundamentals of experimental design for cDNA microarrays. *Nature Genetics* **32**, 490–5.

Churchill GA and Oliver B 2001 Sex, flies and microarrays. *Nature Genetics* **29**, 355–6.

Cleveland WS 1979 Robust locally weighted regression and smoothing scatterplots. *Journal of the American Statistical Association* **74**, 829–36.

Cleveland WS and Devlin SJ 1988 Locally weighted regression: an approach to regression analysis by local fitting. *Journal of the American Statistical Association* **83**, 596–610.

Cobb GW 1998 *Introduction to Design and Analysis of Experiments*. Springer, New York.

Cole ST, Brosch R, Parkhill J, Garnier T, Churcher C, Harris D, Gorden SV, Eiglmeier K, Gas S, Barry CE, Tekaia F, Badcock K, Basham D, Brown D, Chillingworth T, Connor R, Davies R, Devlin K, Feltwell T, Gentles S, Hamlin N, Holroyd S, Hornsby T, Jagels K, Krogh A, McLean J, Moule S, Murphy L, Oliver K, Osborne J, Quail MA, Rajandream M-A, Rogers J, Rutter S, Seeger K, Skelton J, Squares R, Squares S, Sulston JE, Taylor K, Whitehead S, Barrell BG, 1998 Deciphering the biology of mycobacterium tuberculosis from the complete genome sequence. *Nature* **393**(6685), 537–44.

Cormack RM 1971 A review of classification. *JRSS A* **134**(3), 321–67.

Dalgaard P 2002 *Introductory Statistics with R*. Springer, New York.

Datta S and Datta S 2003 Comparisons and validation of statistical clustering techniques for microarray gene expression data. *Bioinformatics* **19**(4), 459–66.

DeRisi JL, Iyer VR and Brown PO 1997 Exploring the metabolic and genetic control of gene expression on a genomic scale. *Science* **278**, 680–6.

Dougherty ER, Barrera J, Brun M, Kim S, Cesar RM, Chen Y, Bittner M and Trent JM 2002 Inference from clustering with application to gene-expression microarrays. *Journal of Computational Biology* **9**(1), 105–26.

Draghici S, Kuklin A, Hoff B and Shams S 2001 Experimental design, analysis of variance and slide quality assessment in gene expression arrays. *Current Opinion in Drug Discovery & Development* **4**(3), 332–7.

Dudoit S and Fridlyand J 2003 Classification in microarray experiments. In *Statistical Analysis of Gene Expression Microarray Data* (ed. Speed T), Interdisciplinary Statistics. Chapman & Hall/CRC, Boca Raton, FL, pp. 93–158.

Dudoit S, Fridlyand J and Speed TP 2000 Comparison of discrimination methods for the classification of tumours using gene expression data. Technical Report 576, University of California, Berkeley, CA.

Dudoit S, Yang Y, Callow M and Speed T 2002 Statistical methods for identifying differentially expressed genes in replicated cDNA microarray experiments. *Statistica Sinica* **12**(1), 111–39.

Efron B 1979 Bootstrap methods: another look at the jackknife. *Annals of Statistics* **7**(1), 1–26. The 1977 Rietz Lecture.

Efron B, Hastie T, Johnstone I and Tibshirani R 2004 Least angle regression. *The Annals of Statistics* **32**(2), 407–99 (with discussion).

Efron B, Tibshirani R, Goss V and Chu G 2000 Microarrays and their use in a comparative experiment. Technical Report 2000-37B/213, Stanford University, Stanford.

Efron B, Tibshirani R, Storey JD and Tusher V 2001 Empirical Bayes analysis of a microarray experiment. *Journal of the American Statistical Association* **96**(454), 1151–60.

Eisen MB 1999 *ScanAlyze, User Manual.* Stanford University, Stanford, http://rana.lbl.gov/manuals/ScanAlyzeDoc.pdf.

Eisen MB, Splellman PT, Brown PO and Botstein D 1998 Cluster analysis and display of genome wide expression patterns. *PNAS* **95**, 14863–8.

Fisher RA and Mackenzie WA 1923 Studies in crop variation. ii. the manurial response of different potato varieties. *Journal of Agricultural Science* **13**, 311–20.

Fraley C and Raftery AE 2002 Model-based clustering, discriminant analysis, and density estimation. *Journal of the American Statistical Association* **97**, 611–31.

Friedman JH 1996 On bias, variance, 0/1-loss, and the curse-of-dimensionality. Technical Report 1996-3, Department of Statistics, Stanford University, Stanford. Available at http://www-stat.stanford.edu/~jnf/?

Galea MH, Blamey RW, Elston CE and Ellis IO 1992 The Nottingham prognostic index in primary breast cancer. *Breast Cancer Research Treatment* **22**, 207–19.

Gasch AP and Eisen MB 2002 Exploring the conditional coregulation of yeast gene expression through fuzzy k-means clustering. *Genome Biology* **3**(11), 0059.1–0059.22.

Ge YC, Dudoit S and Speed TP 2003 Resampling-based multiple testing for microarray data analysis. *Test* **12**(1), 1–77 (with discussion).

Genovese CR and Wasserman L 2002 Operating characteristics and extensions of the false discovery rate procedure. *Journal of the Royal Statistical Society Series B* **64**, 499–518.

Getz G, Levine E and Domany E 2000 Coupled two-way clustering analysis of gene microarray data. *Proceedings of the National Academy of Sciences* **97**(22), 12079–84.

Ghosh D and Chinnaiyan AM 2002 Mixture modelling of gene expression data from microarray experiments. *Bioinformatics* **18**(2), 275–86.

Glasbey C 2001 Image analysis of a genotyping microarray experiment. Technical Report 01-07, Biomathematics and Statistics, Scotland.

Glasbey CA and Ghazal P 2003 Combinatorial image analysis of DNA microarray features. *Bioinformatics* **19**(2), 194–203.

Glonek GFV and Solomon PJ 2004 Factorial and time course designs for cDNA microarray experiments *Biostatistics* (2), 89–111.

Golub TR, Slonim DK, Tamayo P, Huard C, Gaasenbeek M, Mesirov JP, Coller H, Loh ML, Downing JR, Caligiuri MA, Bloomfield CD and Lander ES 1999 Molecular classification of cancer: class discovery and class prediction by gene expression monitoring. *Science* **286**, 531–7.

Hall P and Martin M 1988 On the bootstrap and two sample problems. *Australian Journal of Statistics* **30A**, 179–92.

Hastie T, Tibshirani R and Friedman J 2001 *The Elements of Statistical Learning; Data Mining, Inference and Prediction*, Springer Series in Statistics. Springer-Verlag, New York.

Hastie T, Tibshirani R, Eisen MB, Aliadeh A, Levy R, Staudt L, Chan WC, Botstein D and Brown P 2000 'Gene shaving' as a method for identifying distinct sets of genes with similar expression patterns. *Genome Biology* **1**(2), 1–21.

Hochberg Y 1988 A sharper Bonferroni procedure for multiple tests of significance. *Biometrika* **75**(4), 800–2.

Holm S 1979 A simple sequential rejective multiple test procedure. *Scandinavian Journal of Statistics* **6**, 65–70.

Holter NS, Mitra M, Maritan A, Cieplak M, Banavar JR and Fedoroff NV 2000 Fundamental patterns underlying gene expression profiles: simplicity from complexity. *PNAS* **97**, 8409–14.

Hotelling H 1931 The generalization of student's ratio. *The Annals of Mathematical Statistics* **2**(3), 360–78.

Hubbell E, Lui WM and Mei R 2002 Robust estimators for expression analysis. *Bioinformatics* **18**(12), 1585–92.

Huber W, von Heydebreck A, Sültmann H, Poustka A and Vingron M 2002 Variance stabilization applied to microarray data calibration and to the quantification of differential expression. *Bioinformatics* **18**, S96–S104 (supplement).

Huber W, von Heydebreck A, Seultmann H, Poustka A and Vingron M 2003 Parameter estimation for the calibration and variance stabilization of microarray data. *Statistical Applications in Genetics and Molecular Biology* **2**(1), A3.

Hughes TR, Marton MJ, Jones AR, Roberts CJ, Stoughton R, Armour CD, Bennett HA, Coffey E, Dai H, He YD, Kidd MJ, King AM, Meyer MR, Slade D, Lum PY, Stepaniants SB, Shoemaker DD, Gachotte D, Chakraburtty K, Simon J, Bard M and Friend SH 2000 Functional discovery via a compendium of expression profiles. *Cell* **102**, 109–26.

Ideker T, Thorsson V, Siegel AF and Hood LE 2000 Testing for differentially-expressed genes by maximum-likelihood analysis of microarray data. *Journal of Computational Biology* **7**(6), 805–817.

Irizarry RA, Gautier L and Cope LM 2003a An R package for analyses of Affymetrix oligonucleotide arrays. In *The Analysis of Gene Expression Data* (eds. Parmigiani G, Garrett ES and Irizarry RA), Statistics for Biology and Health. Springer-Verlag, New York, pp. 102–19.

Irizarry RA, Hobbs B, Collin F, Beazer-Barclay YD, Antonellis KJ, Scherf U and Speed TP 2003b Exploration, normalization, and summaries of high density oligonucleotide array probe level data. *Biostatistics* **4**(2), 249–64.

Jeffreys H 1946 An invariance form for the prior probability in estimation problems. *Proceedings of the Royal Society Series A* **186**, 453–61.

Johnson DH and Sinanović S 2001 Symmetrizing the Kullback-Leibler distance. Technical Report 2001-04, Department of Electrical and Computer Engineering, Rice University, Houston, TX.

Kallioniemi A, Kallioniemi OP, Sudar D, Rutovitz D, Gray JW, Waldman F and Pinkel D 1992 Comparative genomic hybridization for molecular cytogenetic analysis of solid tumors. *Science* **258**(5083), 818–21.

Kass RE and Raftery AE 1995 Bayes factors. *Journal of American Statistical Association* **90**, 773–95.

Kaufman L and Rousseeuw PJ 1990 *Finding Groups in Data*. John Wiley & Sons, New York.

Kendziorski CM, Zhang Y, Lan H and Attie AD 2003 The efficiency of mRNA pooling in microarray experiments. *Biostatistics* **4**(3), 465–77.

Kerr M and Churchill GA 2001a Experimental design for gene expression microarrays. *Biostatistics* **2**, 183–201.

Kerr MK and Churchill GA 2001b Bootstrapping cluster analysis: assessing the reliability of conclusions from microarray experiments. *PNAS* **98**(16), 8961–5.

Kerr MK and Churchill GA 2001c Statistical design and the analysis of gene expression microarray data. *Genetical Research* **77**, 123–8.

Kerr MK, Martin M and Churchill GA 2000 Analysis of variance for gene expression microarray data. *Journal of Computational Biology* **7**(6), 819–37.

Khanin R and Wit EC 2004 Design of large time-course microarray experiments with two-channels. Technical Report 04-6, Department of Statistics, University of Glasgow, UK.

Kim H, Zhao B, Snesrud EC, Haas BJ, Town CD and Quackenbush J 2002 Use of rna and genomic DNA references for inferred comparisons in DNA microarray analyses. *Biotechniques* **33**(4), 924–30.

Kirkpatrick Jr S, Gelatt Jr CD and Vecchi MP 1983 Optimization by simulated annealing. *Science* **220**(4598), 671–80.

Kooperberg C, Fazzio TG, Delrow JJ and Tsukiyama T 2002 Improved background correction for spotted DNA microarrays. *Journal of Computational Biology* **9**(1), 55–66.

Lazzeroni L and Owen A 2002 Plaid models for gene expression data. *Statistica Sinica* **12**(1), 61–86.

Lee K, Zhan X, Gao J, Qiu J, Feng Y, Meganathan R, Cohen SN and Georgiou G 2003 Rraa: a protein inhibitor of rnase E activity that globally modulates rna abundance in E. coli. *Cell* **114**, 623–34.

Lee ML, Kuo FC, Whitmore GA and Sklar J 2000 Importance of replication in microarray gene expression studies: statistical methods and evidence from repetitive cDNA hybridizations. *Proceedings of the National Academy of Sciences of the United States of America* **97**(18), 9834–9.

Li C and Wong WH 2000 Model-based analysis of oligonucleotide arrays: expression index computation and outlier detection. *Proceedings of the National Academy of Sciences* **98**(1), 31–6.

Li C and Wong WH 2001 Model-based analysis of oligonucleotide arrays: expression index computation and outlier detection. *Proceedings of the National Academy of Sciences* **98**(1), 31–6.

Li C and Wong WH 2003 DNA-chip analyzer (dchip). In *The Analysis of Gene Expression Data* (eds. Parmigiani G, Garrett ES, Irizarry RA and Zeger SL), Statistics for Biology and Health. Springer, New York, pp. 120–41.

Li C, Tseng GC and Wong WH 2003 Model-based analysis of oligonucleotide arrays and issues in cDNA microarray analysis. In *Statistical Analysis of Gene Expression Microarray Data* (ed. Speed TP), Interdisciplinary Statistics. Chapman & Hall/CRC, Boca Raton, FL, pp. 1–34.

Luan Y and Li H 2003 Clustering of time-course gene expression data using a mixed-effect model with b-splines. *Bioinformatics* 19(4), 474–82.

Maathius FJM, Filatov V, Herzyk P *et al.* 2003 Transcriptome analysis of root transporters reveals participation of multiple gene families in the response to cation stress. *The Plant Journal* 35, 675–92.

McClure JD and Wit EC 2003 Post-normalization quality assessment visualization of microarray data. *Comparative and Functional Genomics* 4(5), 460–67.

McLachlan GJ, Bean RW and Peel D 2002 A mixture model-based approach to clustering of microarray expression data. *Bioinformatics* 18(3), 413–22.

Medvedovic M and Sivaganesan S 2002 Bayesian infinite mixture model based clustering of gene expression profiles. *Bioinformatics* 18(9), 1194–1206.

MGED 2002 Minimum information about a microarray experiment—miame 1.1. Web Publication. Available at www.mged.org.

Michie D, Spiegelhalter D and Taylor C (eds) 1994 *Machine Learning, Neural and Statistical Classification*, Ellis Horwood Series in Artificial Intelligence. Ellis Horwood, Hertfordshire, UK.

Nguyen DV and Rocke DM 2002a Partial least squares proportional hazard regression for application to DNA microarray survival data. *Bioinformatics* 18(12), 1625–32.

Nguyen DV and Rocke DM 2002b Tumour classification by partial least squares using microarray gene expression data. *Bioinformatics* 18(1), 39–50.

Palla L 2004 Bayesian models to test DNA microarray data for differential expression, Universitádi Bologna, Via Belle Arti 41, Bologna, Italy.

Pestova E, Wilber K and King W 2004 Microarray-based CGH in cancer. In *Molecular Diagnosis of Cancer, Methods and Protocols* (eds. Roulston JE and Bartlett JMS), 2nd edn, Vol. 97 of *Methods in Molecular Medicine*. Humana Press, Totowa, NJ, pp. 355–76.

Pollard KS and van der Laan MJ 2002 Resampling-based methods for identification of significant subsets of genes in expression data. Working Paper 121, University of California, Berkeley Division of Biostatistics Working Paper Series.

Radmacher MD, McShane LM and Simon R 2002 A paradigm for class prediction using gene expression profiles. *Computational Biology* 9(3), 505–11.

Ramoni M, Sebastiani P and Kohane IS 2002 Cluster analysis of gene expression dynamics. *Proceedings of the National Academy of Sciences of the United States of America* 99(14), 9121–6.

Raychaudhuri S, Stuart JM and Altman RB 2000 Principal component analysis to summarize microarray experiments: application to sporulation time series. In *BIOCOMPUTING 2000: Proceedings of the Pacific Symposium*, Honolulu, Hawaii (eds. Altman RB, Duncker AK, Hunter L and Lauderdale K), World Scientific publishing Company, Singapore, pp. 455–66.

Ripley BD 1996 *Pattern Recognition and Neural Networks*. Cambridge University Press, New York.

Rocke DM and Durbin B 2001 A model for measurement error for gene expression arrays. *Journal of Computational Biology* 8, 557–69.

Sammon JW 1969 A non-linear mapping for data structure analysis. *IEEE Transactions on Computers* 18, 401–9.

Satterthwaite FE 1946 An approximate distribution of estimates of variance components. *Biometrics Bulletin* **2**(6), 110-114.

Schena M 2003 *Microarray Analysis*. John Wiley & Sons, Hoboken, NJ.

Siddiqui K, Hero A and Siddiqui M 2002 Mathematical morphology applied to spot segmentation and quantification of gene microarray images. *Asilomar Conference on Signals and Systems*, Pacific Grove, CA, http://www.eecs.umich.edu/~hero/Preprints/kashif_asilomar2002.pdf.

Slamon DJ, Godolphin W, Jones LA, Holt JA, Wong SG, Keith DE, Levin WJ, Stuart SG, Udove J and Ullrich A 1989 Studies of the her-2/neu proto-oncogene in human breast and ovarian cancer. *Science* **244**(4905), 707–12.

Smyth GK, Yang YH and Speed T 2002 Statistical issues in cDNA microarray data analysis. In *Functional Genomics: Methods and Protocols* (eds. Brownstein MJ and Khodursky A), Methods in Molecular Biology. Humana Press, Totowa, NJ, 111–36.

Spang R, Zuzan H, West M, Nevins J, Blanchette C and Marks JR 2002 Prediction and uncertainty in the analysis of gene expression profiles. *In Silico Biology* **2**(3), 369–81.

Stein T, Morris JS, Davies CR, Weber-Hall SJ, Duffy MA, Heath VJ, Bell AK, Ferrier RK, Sandilands GP and Gusterson BA 2004 Involution of the mouse mammary gland is associated with an immune cascade and an acute-phase response, involving LBP, CD14 and STAT3. *Breast Cancer Research* **6**, R75–R91.

Stephens M 2000 Dealing with label-switching in mixture models. *Journal of the Royal Statistical Society Series B* **62**(4), 795–809.

Storey JD 2002 A direct approach to false discovery rates. *Journal of the Royal Statistical Society Series B* **64**(3), 479–98.

Suite M 1999 *User's Guide for Use with IPLAB for Macintosh*. Scanalytics Inc., Fairfax, VA. Supplement.

Tamayo P, Slonim D, Mesirov J, Zhu Q, Kitareewan S, Dmitrovsky E, Lander ES and Golub TR 1999 Interpreting patterns of gene expression with self-organizing maps: methods and applications to hematopoietic differentiation. *Proceedings of the National Academy of Sciences of the United States of America* **96**, 2907–12.

Tanner M and Wong W 1987 The calculation of posterior distributions by data augmentation. *Journal of the American Statistical Association* **82**, 528–50.

Tibshirani R 1996 Regression shrinkage and selection via the lasso. *Journal of the Royal Statistical Society Series B* **58**(1), 267–88.

Troyanskaya O, Cantor M, Sherlock G, Brown P, Hastie T, Tibshirani R, Botstein D and Altman R 2001 Missing value estimation methods for DNA microarrays. *Bioinformatics* **17**(6), 763–74.

Tseng GC, Oh MK, Rohlin L, Liao JC and Wong WH 2001 Issues in cDNA microarray analysis: quality filtering, channel normalization, models of variations and assessment of gene effects. *Nucleic Acids Research* **29**(12), 2549–57.

Van 't Veer LJ, Dai H, van de Vijver MJ, He YD, Hart AAM, Mao M, Peterse HL, van der Kooy K, Marton MJ, Witteveen AT, Schreiber GJ, Kerkhoven RM, Roberts C, Bernards PSLR and Friend SH 2002 Gene expression profiling predicts clinical outcome of breast cancer. *Nature* **415**, 530–6.

Venables WN and Ripley BD 1999 *Modern Applied Statistics with S-PLUS*, 3rd edn. Spinger-Verlag, New York.

Wall ME, Rechtsteiner A and Rocha LM 2003 Singular value decomposition and principal component analysis. In *A Practical Approach to Microarray Data Analysis* (eds. Berrar DP, Dubitzky W and Granzow M). Kluwer, Norwell, MA, pp. 91–109.

Weisberg S 1985 *Applied Linear Regression*, 2nd edn. Wiley, New York.

Wernisch L 2002 Can replication save noisy microarray data? *Comparitive and Functional Genomics* **3**, 372–4.

Wernisch L, Kendall SL, Soneji S, Wietzorrek A, Parish T, Hinds J, Butcher PD and Stoker NG 2003 Analysis of whole-genome microarray replicates using mixed models. *Bioinformatics* **19**(1), 53–61.

Wit EC and McClure JD 2003 Statistical adjustment of signal censoring in gene expression experiments. *Bioinformatics* **19**(9), 1055–60.

Wit EC, Nobile A and Khanin R 2004a Simulated annealing near-optimal dual-channel microarray designs. Technical Report 04-7, Department of Statistics, University of Glasgow, UK.

Wit EC, van der Laan M and Palla L 2004b General method for controlling the FDR using mixture models, Technical Report 04-8, Department of Statistics, University of Glasgow, UK.

Witton CJ, Going JJ, Burkhardt H, Vass K, Wit E, Cooke TG, Ruffalo T, Seelig S, King W and Bartlett JMS 2002 The sub-classification of breast cancers using molecular cytogenetic gene chips. *Proceedings of the American Association of Cancer Researchers* **43**(289), 122–8.

Wolfinger RD, Gibson G, Wolfinger ED, Bennett L, Hamadeh H, Bushel P, Afshari C and Paules RS 2001 Assessing gene significance from cDNA microarray expression data via mixed models. *Journal of Computational Biology* **6**(6), 625–37.

Wolkenhauer O, Moeller-Levet C and Sanchez-Cabo F 2002 The curse of normalization. *Comparative and Functional Genomics* **3**(4), 375–9.

Yang YH and Speed T 2002 Design issues for cDNA microarray experiments. *Nature Review: Genetics* **3**, 579–88.

Yang YH and Speed T 2003 Design and analysis of comparative microarray experiments. In *Statistical Analysis of Gene Expression Microarray Data* (ed. Speed T), Interdisciplinary Statistics. Chapman & Hall/CRC, Boca Raton, FL, pp. 35–92.

Yang YH, Buckley MJ and Speed TP 2001 Analysis of cDNA microarray images. *Briefings in Bioinformatics* **2**(4), 341–9.

Yang YH, Buckley MJ, Dudoit S and Speed TP 2000 Comparison of methods for image analysis on cDNA microarray data. Technical Report 584, Department of Statistics, University of California, Berkeley, CA. Available at www.stat.berkeley.edu/users/terry/zarray/Html/image.html.

Yang YH, Dudoit S, Luu P and Speed TP 2002a Normalization for cDNA microarray data. Technical Report 589, Department of Statistics, University of California at Berkley, Berkley, CA.

Yang YH, Dudoit S, Luu P, Lin DM, Peng V, Ngai J and Speed TP 2002b Normalization for cDNA microarray data: a robust composite method addressing single and multiple slide systematic variation. *Nucleic Acids Research* **30**(4), e15.

Yeung KY, Medvedovic M and Bumgarner RE 2003 Clustering gene-expression data with repeated measurements. *Genome Biology* **4**(5), R34.

Zhu T, Chang HS, Schmeits J, Gil P, Shi L, Budworth P, Zou G, Chen X and Wang X 2001 Gene expression microarrays: improvements and applications towards agricultural gene discovery. *JALA* **6**, 95–8.

Index

Statistics for Microarrays: Design, Analysis and Inference E. Wit and J. McClure
© 2004 John Wiley & Sons, Ltd ISBN: 0-470-84993-2